HARP~~I~~

E. Coleman, *ed.*   ~~LECTURE~~ ~~e~~ of TWO CENTS will be
        J. C. Ranso~~m~~

C. G. Jung   PSYCHOLOGICAL REFLECTION~~S.~~
St.-John Perse   SEAMARKS. Translated by Wallace
Jean Seznec   THE SURVIVAL OF THE PAGAN GODS
        Its Place in Renaissance Hu

HARPER TORCHBOOKS / The Academy Library

(continued on next page)

(continued on next page)

1. A queen Polistes wasp depositing an egg in a newly made cell (on right). Other eggs can be seen at the top, larval cells in the middle and pupa cells, closed over, beneath them. A sticky secretion fastens each egg to the side of the cell.

The Social Insects

O.W. Richards

HARPER TORCHBOOKS / *The Science Library*

HARPER & BROTHERS, NEW YORK

THE SOCIAL INSECTS

Printed in the United States of America

This book was first published in 1953 by Macdonald & Co., Ltd., London, and is here reprinted by arrangement.

First HARPER TORCHBOOK edition published 1961

ACKNOWLEDGMENT

It has become very difficult of late for any worker to know all that has been recorded about any group of animals. Even such a restricted subject as social behaviour in insects has by now an immense literature. It is quite impossible for any one worker to have made personal observation on all the species or to have seen all the varieties of behaviour which have been recorded. Any of my entomological colleagues who read this work will be aware how far I have drawn on the writings of naturalists of many countries. I have not thought it necessary to give references to the literature in the text, but I have often mentioned the name of the worker who first recorded some of the more striking facts.

My thanks are due to the University of Hawaii Press for permission to quote the extract from *Insects of Hawaii* by E. C. Zimmermann, Volume V, which appears on page 126.

I am indebted to my colleague Dr. N. Waloff for reading and criticising the first draft of the book.

O.W.R.

London, February, 1953

I have taken the opportunity of the reprinting of THE
SOCIAL INSECTS as a Torchbook to correct a few errors
and add notes on some recent discoveries.

London, October, 1960

O. W. R.

CONTENTS

CONTENTS

LIST OF PLATES

LIST OF PLATES

Acknowledgments for Illustrations

Certain of the drawings in this book are derived from other published books. Acknowledgments are due to Macmillan & Co., Ltd., for Fig. 2 (*Cambridge Natural History* by David Sharpe), Methuen & Co., Ltd., for Fig. 3 (*The Principles of Insect Physiology* by V. B. Wigglesworth), the Author and the Clarendon Press for Figs. 4, 5, 6 (*The Study of Instinct* by Dr. N. Tinbergen), the Author and H.M.S.O. for Fig 7 (*British Bark-Beetles* by J. W. Munro), Cornell University Press for Fig. 8 (*Bees* by von Frisch), Columbia University Press for Fig. 9 (*Ants* by Wheeler), Routledge and Kegan Paul Ltd. for Fig. 10 (*British Ants* by Donisthorpe), Constable & Co., Ltd., for Fig. 12 (*Social Life Among the Insects* by Wheeler).

Thanks are due to the following for permission to use half-tone illustrations: the Director of the Pest Infestation Laboratory, Slough, Bucks, for 19 and 20; Royal Entomological Society of London for 21–26; the Clarendon Press, Oxford, for 42 and 43 (from *The Honeybee* by Dr. C. G. Butler) ; and the Commonwealth Scientific and Industrial Research Organization, Melbourne, for 50 (*Termites [Isoptera] from the Australian Region* by G. F. Hill).

Acknowledgments for Illustrations

Certain of the drawings in this book are derived from other published books. Acknowledgment is due to Macmillan & Co. Ltd. for Fig. 1 (*Cambridge Natural History* by David Sharp); Methuen & Co., Ltd. for Fig. 5 (*The Principles of Insect Physiology* by V. B. Wigglesworth); the Author and the Clarendon Press for Figs. 4, 6 (*The Senses of Insects* by Dr N. Tinbergen); the Author and H. S. O. for Fig. 7 (*Thrips*, Agric. Series by J. W. Munro); Cornell University Press for Fig. 8 (*Bees* by von Frisch); Columbia University Press for Fig. 9 (*Ants* by Wheeler); Routledge and Kegan Paul Ltd. for Fig. 10 (*British Ants* by Donisthorpe); Constable & Co., Ltd. for Fig. 11 (*Social Life among the Insects* by Wheeler).

Thanks are due to the following for permission to use half-tone illustrations: the Director of the Pest Infestation Laboratory, Slough, for 19 and 20; Royal Entomological Society of London for 21–26; the Clarendon Press, Oxford, for 42 and 43 (from *The Honeybee* by Dr C. G. Butler); and the Commonwealth Scientific and Industrial Research Organization, Melbourne, for 20 (*Termites* (*biol*gical) from the *Australian Region* by C. F. Hill.

THE INSECT WORLD

Zoologists have sometimes divided the past history of the world into periods named after the dominant type of animal. The age of fishes was followed by an age of reptiles, and that by an age of mammals. Finally, for the last few thousand years, there has been an age of man. Throughout an immense period, starting before the dominance of the reptiles, insects have been abundant. Though not large and conspicuous enough to give their name to an age, they are at this moment the only group of animals which disputes our dominance. They are infinitely more numerous than the mammals either in species or in individuals; many more specimens of insects can be found in and on an acre of ground than make up the human population of the British Isles. In some parts of the world, biting or disease-carrying insects may temporarily drive man out, and much more often they largely determine the nature of the plant cover and of the sort of animals which live in it.

There is one other respect in which insects seem to rival man: their large and complicated societies are the only ones which can be compared with ours. While they have excited the wonder of mankind at least since the time of Solomon, real understanding of how they work is only just being reached. The ties which unite their societies and the laws which they obey are so different from ours that our emotions are not involved as they are apt to be when we study even the most primitive of human societies. Their most marked characteristic is that within the limits set by the abilities of the species, everything appears to be done for the good of the community and only the necessary minimum for that of the individual. Such societies are well worth studying, not because we are ever likely to wish to imitate them but because of the light they throw on the general principles of organisation and because of the fascination of their extraordinary behaviour. It is, moreover, a field in which many striking discoveries have recently been made, suggesting how much more still awaits the patient investigator.

HOW INSECTS WORK

All the many kinds of insects are built on a rather uniform common plan, very different from the one seen in the mammals. In the diagram of a queen ant on this page, the

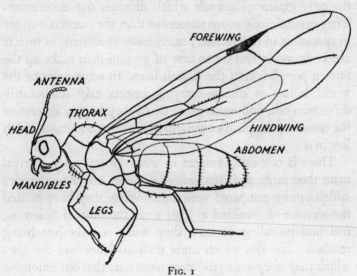

FIG. 1
Queen ant (Lasius), to show the parts of the body.

important parts of the body are indicated. The skeleton of insects is almost all laid down as an external armour. This largely limits their size, for the internal organs of a really large insect would be crushed under their own weight. Moreover, insects breathe by means of a series of fine tubes which ramify into every organ of the body. Oxygen passes in mainly by diffusion and respiratory movements, corresponding to our breathing, are less developed and only aerate the larger tubes. This method of obtaining oxygen is very efficient in small animals but becomes less so as size increases. The blood of insects, also, is not in general enclosed in veins and arteries and does not circulate rapidly round the body like ours, and this again would be a drawback in a large animal.

An insect has a pair of compound eyes made up of a large number of small facets or transparent lenses. The whole eye works on a different principle from ours and there is no way of focusing the vision on objects at different distances. Insects cannot see nearly so far as we can nor form such clear-cut images, but their eyes may be very efficient for certain specialised needs, such as catching quick-moving prey. So much can be deduced from a careful study of the microscopical structure of the eye. The visual powers of an insect like the honey bee can be tested by a process of training. Small glass dishes, one of which contains sugar and water and the others water only, are exposed on small squares of coloured paper. Foraging worker bees can be trained to associate the dish containing sugar with a particular colour. If the dish placed on the colour used for training is replaced by one containing water only, the trained bees continue to visit it for some time, though they cease to feed from it. The postions of the various squares on the table can be frequently altered so that it is clear that the bees are not learning merely the position of the dish. By including amongst the coloured squares all shades of grey, including the whole scale from dark to light, one can show that the bees are responding to colour and not to mere brightness. A colour-blind animal, such as a dog, cannot distinguish a grey square from another colour of similar brightness.

By means of such experiments, it can be shown that the eye of the bee is relatively insensitive to the red end of the spectrum but much more sensitive than our eyes to the ultra-violet end. A curious experiment which demonstrates the same point in ants was carried out by Lord Avebury. He found that if the lid of an artificial nest is made of ordinary glass, the ants endeavour to move into shadow, but if Crookes' glass (which cuts out most of the light at the violet end) is used the ants will stay under it.

Some insects, such as grasshoppers and crickets which produce a lot of sound themselves, have quite elaborate organs of hearing, either in the forelegs or in the abdomen. A female cricket no longer walks towards a singing male if the organs in her forelegs are destroyed, and she can be attracted to a telephone receiver conveying the song of a male from another cage. In other insects, no such organs have as yet been discovered. This is true for instance of the honey bee in spite of the loud hum which it produces. It is possible that some organ is present which has not yet been recognised.

The sense of smell in a honey bee can be investigated by training experiments, similar to those used to study its vision. A number of dishes can be placed in boxes which may be entered through a small hole. The one dish which contains sugar as well as water can have a scent such as lavender water added to it. Foraging bees can then easily be trained to associate sugar with a particular scent. The honey bee, probably because of the importance of flowers in its economy, can distinguish a great variety of scents; it is found to make some of the same errors as we do in confusing certain chemically distinct substances. To us, these substances give out a similar odour. By removing the antennae, it can be shown that the antennae are the chief seat of scent perception. Other insects, such as male moths, are attracted to the scent of the other sex, and will fly to any object with which a virgin female has been in contact.

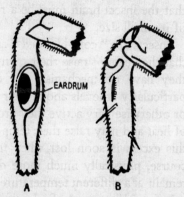

FIG. 2

Two types of ears in the foreleg of grasshoppers, A with the eardrum visible, B with the drum concealed in a slit

The sense of taste has also been studied in some detail in the honey bee by means of training experiments. The sense is much like ours, but its range is a good deal smaller. About one-third of the substances which we call sweet attract bees and some of the things which we call bitter repel them.

An insect's antennae are also important organs of touch and have the faculty of distinguishing hot from cold. If normal honey bee workers are put in a cage of which the floor is hot at one end and cold at the other, most of them collect near the warm end. If the antennae are removed, their position in the cage no longer bears any relation to the temperature of the floor.

In vertebrates, there is a rough correlation between the size of the brain and the mental powers of the animal. Because insects are so small, their brain is minute compared with that of a mammal and on analogy one would not expect it to be very efficient. One certainly would not expect it to be capable of producing intelligent behaviour. It is probable that owing to great differences in plan the analogy is a false one. The behaviour of such insects as the honey bee shows that the insect brain must be a remarkable structure in spite of its small size.

Insects are " cold-blooded " animals. This does not mean that they cannot raise their own temperatures, but only that they have no mechanism for maintaining them constant, particularly at levels above their surroundings. When flying or otherwise very active they produce a considerable amount of heat and may raise their temperatures several degrees, but this excess is soon lost when the activity ceases. It is, of course, physically much more difficult for a small body to remain at a different temperature from its surroundings than it is for a larger one. Scales or hairs may partly hinder the loss of heat but they are not very efficient in small insects. Actually, as will be seen later, most social insects are able to maintain their nests at quite a high and often fairly constant temperature. This is due to the contributions of a large number of separate individuals as well as to the structure of the nest which is usually such as to prevent the flow of heat. Apart, however, from such special cases, insects tend to do everything, including movement, feeding, and development, at a speed which is related to the temperature of their surroundings. Shapley, the American astronomer, measured the speed at which ants travelled along their runs and also the temperature of the soil. Here are some figures for the Argentine ant, each figure being an average of twenty individuals.

| Temperature (°C). | 25.6 | 29.4 | 30.3 | 33.0 |
| Speed of walking (cms/sec.) | 2.62 | 3.49 | 3.57 | 4.17 |

From the speed of walking, the soil temperature can be deduced with an accuracy of one degree. It is probably because of this dependence on external temperature that insects are so much more numerous in the tropics, and that it is there that so much of their evolution has taken place. In many groups, the temperate and still more the arctic species are relatively recent invaders, often with special modifications fitting them to the new conditions. This seems to be true, for example, of our social wasps.

FOOD AND WATER

Insects are not only dependent on the external temperature but also on some external source of water. Active life is possible only if the body fluids are kept at a steady level. Some features of the insects' external skeleton serve to retain water as well as heat; such are the felted hairs which cover some species. But this particular problem is more often solved by modifications of physiology or behaviour. Most species extract all water from their excreta before voiding it, as will be seen later in the termites. Many species do not actually drink water but obtain it from their food. Fraenkel and Blewett showed in the flour moth that caterpillars reared in very dry flour might produce a chrysalis containing more free water than had been present in all the flour which had been eaten. The caterpillars must therefore have combined part of the flour with oxygen to form extra water. This would explain their observation that caterpillars consume more solid food on dry than on damper flour.

The nests of social insects have among their functions that of conserving water, and in artificial nests ants run about as if distracted if they are not kept moist. The termites, because of their soft external skeleton, are the most sensitive of all the social insects to water-loss, and not only are their nests waterproof but termites usually travel everywhere in tunnels which they burrow beneath or construct above the ground.

GROWTH

The location of the insect skeleton on the outside of its principal organs has two important consequences. To get movement, there have to be many joints, and the characteristic insect limb is one result. It is jointed in much the same way as the armoured suit of a deep-sea diver. Again, growth is possible only if the armour is taken off, and this is done in the process of moulting. The external skeleton becomes loosened from the tissues beneath and a new and larger one is laid down beneath it. It is at first soft and wrinkled but very soon after the moult it expands and hardens. In the majority of insects, there is a fixed number of moults during the growing period and they take place at fairly regular intervals. Once the adult form is reached, no more moulting occurs and there is no more growth. Thus one can roughly estimate the age of a honey bee worker by the amount of wear which its wings and skeleton show.

The physiology of growth and of moulting in insects has been very actively studied during the last twenty years. The story is complicated and not yet fully understood, but it seems that moulting and the changes of form to which it gives rise are controlled by the brain and by certain ductless glands, mostly associated closely with it. In the bug Rhodnius Wigglesworth showed that the early stages do not moult if decapitated during the first five days after the large meal of blood which normally suffices for the whole period. Bugs decapitated later moult normally. This suggested that a substance was being produced in or near the head; when enough had been accumulated in the blood, the insect could moult even if the head was removed. This interpretation was confirmed by decapitating two bugs, one which had moulted less than five days previously and one which had passed this critical period. The two headless insects were connected by a fine glass capillary tube and it was then found that they both moulted at the same time. This control of form and structure,

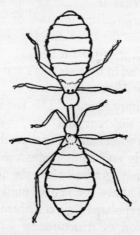

FIG. 3

Two 4th instar nymphs of Rhodnius decapitated and connected by a capillary tube sealed into the neck with paraffin wax (after Wigglesworth).

partly nervous and partly glandular, may, it is beginning to appear, have great significance in some social insects. In termites especially, the body sometimes seems to be moulded by the needs of the community, and the intervention of the brain may make this possible.

REPRODUCTION

In most insect species, the mature female, after being fertilized by one or more males, lays between fifty and one thousand eggs; the actual number largely depends on the manner of life of the species. These eggs hatch out into a young animal which in the less specialized insects is termed a nymph. This differs from the adult chiefly in its smaller size and in its undeveloped wings and reproductive organs. Amongst social insects, this type of metamorphosis occurs among the termites, sometimes called white ants. In other insects, there has been a gradual divergence between the early and adult stages, so that one gets for instance the familiar contrast between the caterpillar and the butterfly. What hatches from the egg is a larva which lacks many of the characteristic features of the adult. It often happens that when the larva develops in a protected situation with a rich source of food it becomes a legless maggot or grub. This is found in the ants, bees and wasps, but the grub is not especially associated with social life, since it is found equally in solitary species. The larval and adult stages are specialized for different functions, the former for feeding and growth, the adult mainly for reproduction. In the simpler insects, the early and adult stages are still sufficiently alike for one to change gradually into the other. In the more complex insects, they have become so unlike that a resting stage, or pupa, is interpolated between them, to allow of extensive reorganisation of structure.

Although it was stated earlier that the number of moults is usually fixed, there are many exceptions. Thus the larva of the carpet beetle can stand weeks or months of starvation. If it is not feeding, it gets smaller at each moult. By a process of alternate starvation and feeding, the American entomologist Wodsedalek got one of these larvae to live for five years, getting first smaller, then larger. This potentiality for variation in the number of moults is of great consequence in the social life of termites, though it is not in them known to be controlled by nutrition.

BEHAVIOUR

Probably insects appeal to most of us in the first place through our sense of beauty and strangeness. Most of us have seen a red admiral on the Buddleia in our garden or an eyed hawk moth on a paling. If we probe more deeply, the story of insect behaviour can be just as engrossing. It is soon apparent that many of their actions have, in some sense, a purpose, but if we are not careful every word we use carries into our discussions implications which it has acquired in human intercourse. The red admiral feeds on the nectar of Buddleia, but the female goes to nettle to lay her eggs and it is on nettle that the caterpillar feeds. The behaviour of the butterfly has a purpose in the survival of the species, but it is in the highest degree improbable that the female knows what she is doing; she has no purpose in the human sense. Her behaviour is instinctive: that is, it is not learnt but produced automatically at the appropriate time. We have no way of knowing what an insect is thinking nor whether it is conscious at all, and it is best to avoid, as far as possible, language which implies that we do know. When we poke a wasps' nest with a stick, the " angry " wasps buzz out and attack us. We use the word angry to avoid a clumsy circumlocution, but we should not suppose that the wasp feels as we do when our house is broken into by a burglar. We cannot tell what the wasp feels, we can only see how it acts. If sometimes in describing social insects it is convenient to use words which have human applications, this is only for convenience and in order to be more concise.

The simplest type of insect behaviour consists in immediate reactions to messages received from the outside world—the rays of the sun, the water vapour emanating from damp soil, the smell or appearance of food or of the other sex. It is an

25

immediate reaction of this type which leads a grasshopper to sit on the sunny side of a mound and to place itself so that the largest possible area of its body is exposed to the sun. Similarly the fully grown maggots of blowflies crawl away from the light, and in nature this leads them away from their food to pupate in the soil. In an often-repeated experiment, the maggots are placed on a board which is lighted from two sides by beams of unequal intensity. The track of the maggot then divides the angle between its original position and the lights in proportion to their brightness.

Another simple type of behaviour is the reaction to internal messages, such as those from the stomach or alimentary canal which lead an insect to search for food, or from the ripening ovaries which lead to a search for an egg-laying site. A hungry insect may become restless and merely wander about, more or less at random, until it finds food. External stimuli such as light, shade, and wind, will partly determine its path. If normal spider beetles are put in a cage of which one end is kept moist, the other dry, most of them will collect in the dry end; but if they have been kept very dry for some days previously, they collect at the damp end. Their behaviour is a resultant between an external gradient and their internal condition. Behaviour of this sort, where the pattern is simple, is not usually called instinctive. Where the behaviour is more elaborate, as when the red admiral seeks out a nettle and lays eggs in a characteristic way on a special part of the plant, the pattern is said to form an instinct. Instinctive behaviour has a complex, inherited pattern. The performance is normally perfect on the first occasion, so that learning is not involved. The sequence of acts requires a particular internal condition of the animal, but it can be carried to completion only in a situation which also provides certain specific external stimuli.

The hunting behaviour of the solitary wasp, Philanthus, described by Tinbergen, is a good example of a complex, inherited pattern. A hunting female at first uses her eyes and examines any object of the right size, hovering in the air 4–6 inches away. An attack is only made if the object has

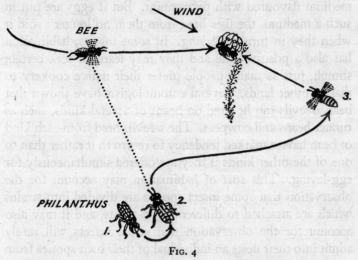

FIG. 4

Hunting behaviour of the solitary wasp, Philanthus. A hunting female at first uses her eyes and turns towards and examines any object of the right size, hovering in the air four to six inches away. An attack is only made if the object has the smell of a honeybee which is the normal prey (after Tinbergen).

the smell of a honeybee, which is the usual prey. The attack is usually made upwind, along the gradient of the smell.

Besides following their instinctive behaviour patterns, all insects are to some extent influenced by their past experience, that is they have some power of learning. Learning is of more than one kind and the simpler types are probably possible for most species, though only a few examples have been well investigated. One sort which may be universal is called habituation. Most animals which avoid noise and vibration can learn that some regularly recurring noise is harmless to them, just as we acquire the power of sleeping through the noise of a nearby railway. Thorpe showed that the vinegar fly will not normally lay its eggs in an agar medium flavoured with peppermint. But if eggs are put in such a medium, the flies bred from them no longer avoid it when they in turn are laying. In some insects, habituation has also a positive side and they may learn to seek certain stimuli, just as many people prefer their native cookery to that of other lands. Several entomologists have shown that bean weevils can be bred on beans of several kinds, such as runner beans and cowpeas. The weevils bred from each kind of bean have a marked tendency to return to it rather than to one of the other kinds if they are offered simultaneously for egg-laying. This sort of habituation may account for the observation that some insect species are divided into strains which are attached to different food-plants, and it may also account for the observation that social insects will rarely admit into their nests an individual of their own species from another colony. The stranger evidently does not have the right smell, though sometimes he may be able to acquire it.

A more complex type of learning which is also very widely spread is known as trial and error learning. This is demonstrated clearly when for example a cockroach is placed in a maze containing food. The maze is a box divided by barriers into a number of passages, some of which are blind alleys whereas others eventually lead to the food-chamber. It is arranged so that the correct turning is sometimes to the right, sometimes to the left, and the time which it takes for the cockroach to find its way through the maze to the food can be recorded. In these experiments, the insect, just as a man might, takes at first many wrong turnings, but gradually the correct path is learnt. A good deal of insect behaviour could be of this type; but unless conditions are controlled so that it is known what stimuli are being received, it is impossible to be sure that the insect is not making a series of instantaneous reactions. It might, for instance, reach a source of food whose smell became stronger and stronger the more nearly it was approached, without any need to learn the way.

FINDING THE WAY

A more complex type of behaviour is described as insight learning. What this means can be explained by an example, though there is still some dispute whether any insect is capable of it. (In my own view, it is no longer possible to deny some power of insight learning to some of the Hymenoptera, including not only social forms like ants, but also the solitary, nest-building bees and wasps). All bees and wasps make a careful study of the nest-site, walking or flying round it in a number of widening spirals or circles. It is often possible to prove that they are memorising small local landmarks.

Thus, Tinbergen describes how a circle of pine-cones was made round the burrow of the wasp, Philanthus, when the wasp was inside. When she came out, she flew round studying the site for six seconds. While the wasp was away hunting the cones were moved to form a circle one foot away from the nest. When the wasp returned after ninety minutes with a captured honeybee, she looked for the nest in the circle of pine-cones and in no case found the entrance until the cones were again placed round it. Experiments with ants, to be described later, show that both the position of the sun in the sky, and also a scent-trail, may help to indicate a path.

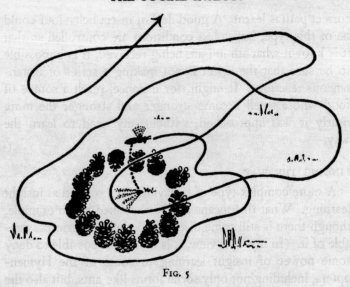

FIG. 5

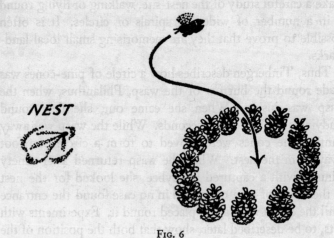

FIG. 6

How the alteration of landmarks confuses a female Philanthus
when returning to her nest with prey (after Tinbergen).

The evidence for insight learning is based on more elaborate behaviour of this type. A solitary spider-hunting wasp, Anoplius fuscus, has been much studied. When it has stung and paralysed its prey, it hides it in a tuft of vegetation and, after making a short locality-study on foot, searches for a place to dig its nest. Once when the nest was several feet away from the spider, the Swedish entomologist Adlerz put a white towel, of a size 2 x 3½ feet, on the ground between the wasp's nest and the spider. The wasp normally visits the spider several times while the nest is being constructed. On such a visit, when she came to the towel she was completely nonplussed. She would only walk an inch or two on to it and, eventually, after casting round for half an hour, she found her way round the edge of the towel and ran to the spider. Coming back to the nest, she took a bee-line across the towel and later the spider was dragged to the nest by the same path. This suggests that the wasp appreciates the relative positions of the spider and the nest and that once the intruding object has been studied she can pass from one to the other without using detailed land marks.

Still clearer evidence for insight learning has been found by Thorpe in the solitary wasp, Ammophila pubescens, which, as we shall see, is known to be highly endowed in other respects. This wasp catches caterpillars and drags them back to her nest. When metal screens were put in her path she merely made a detour round them and then went straight on in the right direction. Even if the wasp was caught and carried some distance in a dark box, she was usually able, when released, to set off straight for her nest. She gave the impression of knowing all the landmarks on a considerable area of rough ground so well that she could quickly find her nest from any starting point.

The knowledge of the relative positions of numerous land-marks is something different from being able to follow a single beaten trail. Social wasps, also, learn to know a large district surrounding their nest. An experiment by Gaul in the U.S.A. provides a curious commentary on this fact. He brought a nest from a mile away and put it in a hive. The adult wasps all deserted their nest and brood and were found building a new nest at a few feet from the original site. They will only stay in their nest in the hive if they are shut up in it for the first eight hours.

As will be described later the honey bee has the further power of communicating to its nest-mates information about the location of sources of food. Most social insects, also, have an advantage over most solitary ones in having a less rigid type of behaviour. In many solitary wasps the successive phases of nest-construction follow one another in a regular sequence which does not allow for unforeseen accidents. In Fabre's study of the wasp, Pelopaeus (now called Sceli-phron), he found that when the mud-cells had been stored with spiders, the whole group of cells was covered over with an irregular daub of mud. If he removed the cells, the wasp still daubed over the place where they had been. This irra-tional rigidity is characteristic of all instinctive behaviour, but is especially marked in solitary species, though there are exceptions. No such complete rigidity is possible in social wasps. The nest is growing all the time, and many different jobs must be done in an order which depends on circumstances. The wasps may have to fetch building material, or water, or food, or add cells to the nest, or feed the brood. Thus each individual worker has to do a greater variety of things than any solitary species and must be prepared to do them in a different order each day, since specialisation for particular tasks is little developed in the wasps.

2. A mud-dauber wasp (Sceliphron caementarium). The female captures spiders and paralyses or kills them by stinging. She stores them in mud cells, laying one egg on each cell-full, thus supplying food for the grubs.

3. A sand wasp (Ammophila sabulosa) dragging a caterpillar to its burrow.

4. Hunting wasp, Ammophila sabulosa.

5. Another hunting wasp, Sphex maxillosus.

6 and 7. Young nests of the German wasp (Vespula germanica), before any workers have hatched. Part of the envelopes of the nest on the right have been removed to show the comb.

HOW SOCIAL BEHAVIOUR DEVELOPS

We may now consider what we mean by the term " social insect ". Many solitary insects are gregarious, that is they share certain common needs or react in the same way to certain external stimuli so that dense populations assemble locally. Many unrelated species of caterpillar may be found feeding on one plant, and large numbers of assorted insects may be trapped at a street-lamp. A suitable earthbank may harbour the nests of hundreds of solitary bees or wasps, but they never help one another, and indeed frequently fight, like two families trying to share one kitchen. Gregariousness has never led to social life in insects, though in the locusts, where dense populations of nymphs aggregate into marching bands and where the adults may fly in enormous swarms, there is at least a rudimentary organisation of the behaviour of the group.

All animals must behave so that their young will find the conditions necessary for survival and growth. In the swift moth this instinctive maternal care amounts to no more than scattering eggs broadcast over a field, where there is a good chance that the young caterpillars will be able to burrow into the soil and to find the roots on which they feed. In this moth, as in most other insects, the female also requires the co-operation of a male for a brief period, to fertilise her eggs. The female of the large cabbage white butterfly lays her eggs in a group on a cabbage, the food-plant of the caterpillar. These when young sit side by side on the leaf, and it is thought that this gives them some mutual protection owing to their unpleasant smell and conspicuous colour. The same sort of behaviour is seen in the cinnabar caterpillar on ragwort. Other caterpillars, like those of the small ermine moth on the garden euonymus, spin a joint web in which a number hatched from one egg-group hide together by day and from which they come out to feed by night. Any advantage acquired by such association is essentially a social one, but their behaviour has never become very elaborate. A caterpillar is a relatively simple insect compared with, say, an ant, and is in any case only a transitory stage.

Nearly all the advanced insect societies appear to have arisen from a progressive development of maternal care, the social unit being the mother and her offspring rather than a collection of brothers and sisters. The first essential step is that the female should continue to have contact with her eggs after she has laid them, and usually after they have hatched. It is impossible even to list the immense variety of ways in which insects have improved on the simple act of laying their eggs on or near the larval food-supply. Some ichneumon wasps, instead of laying their eggs straight into the body of a caterpillar, carry them about hanging from the ovipositor until they hatch. Only then do they put the young grubs on to the sawfly larva which forms their food. Though this particular type of behaviour seems to have been a blind alley, anything which prolongs the contact between the mother and her young is a start on a possible road to social life. A number of plant-feeding or sometimes predatory bugs brood over their eggs and younger nymphs. The nymphs tend to form a compact group and the female stands over them when anything approaches, rather like a hen brooding over her chickens. This example is also comparable with the cinnabar caterpillars already mentioned, since the bugs are brightly coloured and have an evil smell.

A group of insects, known as the Embiids, which are found in most of the warmer parts of the world, are often cited as an example of rudimentary social behaviour. They are also of interest because they seem to be allied to the termites, in which the social life of insects reaches one of its peaks. The Embiids are able to spin silk which comes as a viscous thread from hollow hairs in their forefeet. They weave irregular chambers on or under the bark of trees and there live together in groups of some size. The group includes the developing young as well as a number of males and females. The chief advantage is probably the protection given by the communal web, since each individual finds its own food.

Other examples of maternal care which has developed into rudimentary social behaviour are found in the earwigs and in various beetles. The common earwig lays her eggs in a small hole which she excavates in the ground. She sits with them and turns them over occasionally and licks them. When the mother is removed, the eggs always seem to die, probably because a fungus grows on them when they are not cared for. The young stay with the mother only for a few days after hatching. Hinton has shown that one of the common British rove beetles has rather similar behaviour. The female excavates a chamber in a pat of cow-dung and lays her eggs in it. She will defend them against intruders, including the larger larvae of her own species. When the eggs hatch, the young larvae stay with the female until just before their first moult.

Rather more elaborate social behaviour, in which the males also share, is seen in several other kinds of beetle. The dung beetles, some of whose habits were described by Fabre in his book on the " Sacred Beetle ", are an example. In Geotrupes typhoeus, which is found on some of the commons round London, the male and female co-operate in digging a deep burrow and in storing dung as food for the larvae. The female, at any rate, stays with the young for most of their development. In some of the other kinds of dung beetle the male and female also co-operate in rolling pellets of dung to the mouth of the burrow, and both parents may stay in the burrow until the young beetles of the next generation are ready to leave it. In the burying beetles a male and female co-operate in burying small dead animals by digging away the earth from underneath them. The pair of beetles then

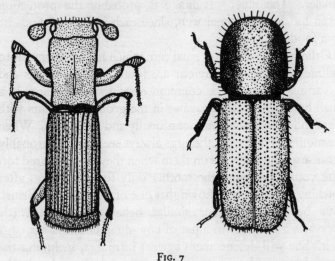

FIG. 7

Left, an ambrosia beetle, Platypus cylindrus. Right, a bark beetle, Xyloborus saxeseni. (Redrawn from Munro)

live in a chamber surrounding the mass of carrion. The female lays her eggs in the earth, in the walls of a tunnel leading out of the chamber. After about five days, the young larvae hatch and burrow through the soil to reach the chamber. Here, up to the first moult, and for a short period after each succeeding moult, they are fed by the female on regurgitated food. It has been found that though some of them may develop without this attention a much smaller proportion do so successfully.

Two allied families of beetles, the bark beetles and the ambrosia beetles, though very different from the kinds which feed on dung or carrion, have developed essentially similar habits. The females, in some species assisted by the males, excavate a tunnel in solid wood. A special fungus grows in the tunnel, its spores being carried there either in a hollow in the head of the female or in her alimentary canal, an interesting analogy with what happens in the fungus-culti-vating ants and termites, described later. The female lays her eggs in side chambers and the young larvae may be fed by the female on pieces of the fungus. In some species, the male may guard the main entrance, perhaps chiefly to keep out other males. These are remarkable societies while they last, but they are never prolonged beyond the one generation.

FULLY SOCIAL BEHAVIOUR

It appears that to obtain a larger and more elaborate social organisation the female must live long enough to overlap with several generations of her young. Some solitary insects do live long enough for this to be possible, though in most of them there is very little overlap between the lives of the adults of successive generations. An even more important step is for more than one female to co-operate in looking after the young. This happens in none of the elementary social groups; it provides the best distinction between social and sub-social insects. A true social insect may be defined as one in which the female tends or helps to construct a brood-chamber for an egg (or larva) laid by another female. This condition is realised only in the ants, bees, and wasps, belonging to the order Hymenoptera, and in the termites; the latter are unlike the Hymenoptera in that the males play as big a part in the colony as the females. Only a very few insect species have been able to develop fully social habits.

Of the 20,200 species of insects in the British fauna, only thirty-seven ants, about forty-four bees, and seven wasps are social. The proportions in other parts of the world are probably much the same.

The insect society, with its store of food for the young, is bound to provide a tempting site for a parasite to lay its eggs. Moreover, as explained in Chapter 7, other females of the same or allied species will often attempt to usurp the incipient colony. Thus a certain "jealousy" of oviposition by other females is nearly always shown by the female founder of the colony. In humble bees and in most kinds of social wasps the queen attacks any other individual attempting to lay eggs, and this behaviour is aroused much more by the act of oviposition than by the mere presence of another fertile female. It is also true that if many females all lived together and laid eggs indiscriminately, the number of young would soon be too great for the available food-supply and for the number of nurses. In all social forms a distinction, not always very sharp, has thus developed between egg-laying queens and sterile workers. Even if the workers do, in some circumstances, lay eggs, these are not fertilised and normally produce males only.

It appears that the special form of sex-determination which is found in most Hymenoptera made the attainment of social life more easy in that group. The female can store up internally for long periods the sperm transferred to her by the male; and she is able to expose some of her eggs to sperm just before they are laid, whereas others can be laid unfertilised. In nearly all Hymenoptera the fertilised eggs produce females and the unfertilised ones males. Thus for the maintenance of the colony males are needed only at long intervals, when a new brood of young queens requires fertilisation. Among the stimuli which determine whether the queen lays fertilised or unfertilised eggs are probably some depending on the general condition of the colony which can thus adjust the sex-ratio to current needs.

In the typical Hymenopterous colony no males are present for most of the year : the workers are sterile females and no corresponding caste of sterile males has been evolved. The termites, in contrast, have both sexes equally represented in the sterile castes. In this group the female seems to be unable to store for long periods the sperm received at her first marriage flight, and she therefore requires frequent fertilisation. Moreover, the method of sex-determination is different : unfertilised eggs do not develop, while fertilised ones give rise to about equal numbers of the two sexes. It may be that the difficulty of evolving two sterile castes is one reason why the bisexual type of society arose only once, in the termites, whereas the female-dominated type evidently arose four or five times in different groups of Hymenoptera.

2

SOLITARY AND SOCIAL WASPS

Among the social insects the wasps seem to have changed least from their solitary ancestors. Their colonies, even when large and elaborate, never seem to attain to the high organisation of the beehive or the ants' nest. Moreover, there are good grounds for thinking that both bees and ants were evolved in the remote past from ancestral wasp-like creatures. The potentialities of the wasp-stock were probably all present in the ancestor of the two other groups.

A HUNTING WASP

Even a solitary wasp may be capable of elaborate behaviour, and there is no better example of this than the species Ammophila pubescens, which has become famous in the last few years and is found in southern England. Many observers had built up an incomplete picture of its behaviour over the last fifty years, but during the war a very long and detailed study of its habits was made by Baerends in Holland. He marked many females with coloured paints so that each one could be recognised again. In this way a detailed picture of the nesting-cycle could be constructed. Soon after fertilisation, the female digs a nest in the form of a burrow in sandy soil. The gallery descends for about an inch and then bends at right angles and ends in a small oval chamber. Baerends removed such nests bodily from the soil and replaced them with a split block of plaster of Paris in which an artificial nest had been cast. Such a block when dusted over with sand was accepted by the wasp, and had the advantage over the real nest that its contents could be examined without doing much damage to the site.

After making the nest, the female catches a caterpillar, stings it until it is paralysed, drags it into the nest, and lays an egg on it. The nest is closed, and opened again only after some days when the wasp-grub will have hatched and may need a second caterpillar. Adlerz in Sweden had suggested that each female might be able to look after more than one nest at the same time, and this Baerends found to be true. Sometimes as many as three nests were kept going, each in a different stage and needing different treatment. In one the egg might not have hatched, in the second the grub might need more food, whereas the third might be ready for final closure. The female at first sight gives the impression of knowing what treatment each nest requires, but experiment showed that the " knowledge " was of a very primitive type. Normally, the wasp makes a morning inspection of the nests and this determines how she treats each one for the rest of the day. Thus if on the first inspection she finds a nest needs food, food is put into it even if a caterpillar has already been placed experimentally in the nest. She seems to be unable to revise her first impression until another twenty-four hours have passed. Nevertheless the successful accomplishment of several elaborate duty-sequences within one day, together with her marked ability in finding her way about, make this wasp one of the most gifted of the solitary species.

THE VARIETIES OF WASP

In Britain the word " wasp " is usually applied to one of the common social species which eat fruit and sometimes

interfere with picnics. There are many other insects, including even moths, with a similar black and yellow livery,
which are often mistaken for wasps and may indeed
gain some protective advantage if such mistakes are made
by birds.

Entomologists, however, apply the word wasp in the broad
sense to several large groups of insects, many of which do
not show the black and yellow colours. The ichneumon
wasps, for instance, lay their eggs in or on other insects,
more rarely in spiders, and other animals. The act of laying
the egg seems to disturb the host very little, and it may
survive for a long time. But in the end, after the egg has
hatched, the ichneumon grub eats the entire contents of its
host, so that nothing but an empty skin is left. An insect
which does this is usually called a parasite; but it can just as
well be called a specialised predator which kills its prey at a
convenient moment.

Solitary hunting wasps paralyse or kill their prey (which
may be a caterpillar, spider or something else of that sort) and
store it in a nest. Here again the prey is eaten by the grub
which hatches from the egg which the wasp has laid in the
nest. This type of behaviour is associated with a change in
the structure of the wasp's ovipositor: it becomes a *sting*. In
the ichneumon, the egg can be passed down between the
stylet-like blades of an ovipositor which thus acts like a
hypodermic syringe for injecting the egg into the host. In
the hunting wasp an egg pore is found at the base of sting;
the ovipositor is then only a syringe for injecting poison
into the prey, or into any marauding enemies. A sting is
found in the nest-making ants, bees and wasps, though in
some of them it is not actually syringe-like.

The solitary wasps themselves are a very large group
with a great diversity of appearance and habit. Some of

them behave in almost the same way as the ichneumon. The Scoliid wasps, which are common in most of the warmer parts of the world, attack the underground grubs of cock-chaferlike beetles. The wasp has powerful spiny legs by which she can work her way underground to find her prey. Some species sting the grub and lay an egg on it without making any nest, leaving the grub in its own burrow. Others do make a small underground cell to which they drag the grub before laying on it. Most of the solitary wasps, however, make a nest, usually a gallery ending in one or more cells either in the soil or in rotten wood. A few of them can build beautiful mud cells, as described for instance by Fabre in Pelopaeus (Sceliphron) ; less commonly they build cells of vegetable matter such as leaves or resin.

The nest-making solitary wasps fall into three principal groups which have a different pattern of behaviour. The least advanced type are the spider-hunters which first catch and paralyse a spider, then hide it and look for a nesting-site. Usually, they make only one cell at each place, and when they have stored the spider in it wander off and repeat the process. This pattern is not very different from that of Scolia, and only very few of the spider-hunters make mud-nests or construct several cells in one place. This means that the position of the single cell has only to be remembered for a short time, at most for a day or so, after which it is visited no more.

The sand wasps always make a nest before catching their prey; this may be any sort of insect or spider. The nest consists of several cells which are typically sub-divisions or branches, often slightly enlarged, of a long tunnel. When several prey are stored in the cell, the egg is normally laid on the last one brought in and the cell is then closed. Thus the mother has no contact with her young, but she must become very familiar with the surroundings of the nest since many, sometimes all, her cells are in the same spot.

A third group, the Eumenids, resemble the social wasps in many details of structure, but build nests very like those of the sand wasps. They have one habit which has been retained in the social species : the egg is laid in an empty cell and the prey, which are paralysed caterpillars or beetle grubs, are brought in afterwards.

These methods of providing for the young are known as mass-provisioning, since the egg is enclosed in a cell with enough food for it to complete its development. The food is supplied whole, in the form of paralysed or sometimes dead insects. A few species both of sand wasps and of the Eumenid wasps show an important advance in behaviour ; they practise progressive provisioning. These wasps lay an egg either on the first prey brought in or in the empty cell and then wait until the egg hatches before bringing more food. In this way the mother has some contact with her young. Ammophila pubescens, already described, provides one of the best known examples of such behaviour. In some of the other species which practise progressive provisioning, such as the African wasp Synagris, the female may feed her young with chewed fragments of prey instead of whole specimens.

SOCIAL WASPS

It is from wasps with this sort of equipment, though not from any living solitary species, that we imagine that the social species arose. The ancestors of the social wasps were thus sting-bearing hunters, capable of building nests of some complexity, capable of learning their location with great exactness, and beginning to have some contact with their young. However, in the solitary species each nest is made

by a single female and the individuals are either males or fully
fertile females. In the social species, at least some of the
females are relatively or quite infertile and these assist in the
care of the young of the fully fertile females, or queens. The
females with reduced fertility are almost always quite incapable of fertilisation by the males and form the worker caste.
A further feature in which the solitary wasps differ from the
social ones is that the female rarely lives much longer
than the male, and successive generations of adults hardly
overlap.

The most familiar social wasps in Britain and North
America are species of the genus Vespula. These will be
described first, although some of the less familiar tropical
species have simpler behaviour and seem to come nearer to
the solitary species. These common wasps have two sorts
of females, the egg-laying queens and the workers. The latter
are considerably smaller than the queens, have a slightly
different pattern of yellow markings, and are not fertilised.
Males occur for a rather short period at the end of the summer, when they fertilise the young queens of the next generation. The workers cannot correctly be described as sterile,
since some of them may lay eggs, especially towards the end
of the season or even sooner if the queen dies; but these
unfertilised eggs produce only males, so that the workers
alone could not carry on the species.

The young fertilised queens may stay for a little time in the
nest where they were born, but quite soon they begin searching for winter quarters, since they are the only members of
the colony which hibernate. At the first frosts or earlier, the
males and workers die off and all that is left is the young
queens hiding in the ground, under dead leaves, or under
bark. A few of them may try to hide indoors, but it is usually
too dry for them there and the warmth prevents them from

resting undisturbed until the next spring. Under natural conditions they pass into complete immobility with wings and legs folded up, and they do not begin to move again until March or April. Almost at once, on becoming active they begin to look for a nesting-site. In the two commonest English species, the common wasp and the German wasp, this is usually a hole underground, often an old rat- or mouse-hole. Here the queen builds her first nest. This, like those of all the social wasps, is made of a sort of paper. The paper is formed from wood fibres scraped from posts, palings, stems of dry hogweed or rotten wood. The marks of wasps' mandibles can be seen on almost any unpainted wood fence in England, forming little scratches often tending to run into lines. The wood fibres are chewed up with saliva to make a paste which when spread out thin dries into a grey or brownish paper. The common wasp uses mainly rotten wood and its paper is yellower, softer and more brittle; the German wasp, so-called because the eighteenth-century entomologists received specimens from that country, uses sound wood, and its paper is greyer and tougher.

The beginning of the nest is a little pillar, hanging often from a root which projects downwards from the roof of the hole. At the end of the pillar (which is about half an inch long) is built a cell, and round the sides of the first one others are added. The beginning of the first cell is a hemisphere, hanging upside down; the second cell starts as a sort of crescent-shaped pocket on one side of the first. Later cells are added in the angle between two earlier cells. When a few cells have been started the queen begins building *envelopes*. These are dome-shaped covers attached near the point from which the pillar hangs or, in the later envelopes, to an earlier one. They form cup-like domes over the cells and later completely enclose them except for a small more or less circular entrance at the bottom. The construction of several (two to four) envelopes with air in between provides the best sort of heat-insulation that can be imagined. Eggs are laid in cells which have only been started, but when the grub hatches and begins to grow the cells are lengthened to keep pace with it. The first cell remains circular in cross-section, but the later ones become more and more regularly hexagonal, each of their walls, except those on the periphery of the comb, being shared with another cell. It can be shown that this arrangement makes possible the most economical use of a given quantity of paper. It seems that as wasps do not vary much in size, they can use their own bodies as a pair of dividers for measuring a standard distance.

The queen wasp never makes more than about ten to twenty cells, since when the grubs are growing actively it is all that she can do to keep them fed unaided. It seems clear that adult wasps can take only liquid foods, since the entrance to their throat is too small to admit more than the minutest solid particles. Probably most of their food is nectar from flowers, juices of fruit and sweet things of that sort. But to feed their young they catch and cut up other insects, such as flies or moths. These are pounced on when settled. They are not stung but cut up and chewed into a paste. The wasps will also cut small pieces from dead animals. Probably some of the juices extracted from the paste form a necessary supplement to their own diet. Whether the young receive much sweet stuff as well as animal matter is uncertain. The grubs live and grow upside down, since the opening of the cell is at the bottom. When full-grown the grub spins a silken cocoon which forms a white cap across the bottom of the cell and a thin lining to the rest of it.

The life of the queen with her first cells differs in no important respect from that of a solitary wasp practising progressive provisioning, but when the first workers bite their way out through the cocoon a real social unit has arrived. With the help of the workers the comb begins to grow rapidly, and a new one will soon be started. To contain the combs, the envelopes will be reconstructed: the paper is chewed away and used with new material to make bigger ones. The queen gradually stops leaving the nest and soon performs no duty beyond egg-laying. Later in her life, when she has been laying eggs rapidly for a long time, her abdomen takes on a discoloured greasy look and she loses all power of flight, so that the old queen is easy to recognise.

The second comb starts as a pillar hanging from the centre of the first comb, and the process of growth is much the same as in the first; but large combs are always supported from the one above by several pillars, and their edges are attached to the envelope by small lateral struts. The workers get from one comb to the next by crawling round the edge between the comb and the envelope. As the nest grows, it is nearly always necessary for the wasps to enlarge the hole in which it is hanging. They do this by bringing in water to soften the earth, and in a populous nest workers can often be seen flying out one behind the other carrying little pellets of mud. A big root or stone which they cannot remove may cause the nest to acquire an irregular shape.

As the comb starts from a cell near its centre the first eggs are laid here too, and as a result concentric rings of developmental stages are often seen in a comb. The centre may consist of cocoons, surrounded by rings of grubs getting progressively smaller towards the periphery, till finally there are rings of cells either with eggs or not yet occupied. However, after a new worker has bitten its way out of the cocoon, the cell is cleaned up and used again, so that in a somewhat older comb the central cells will contain eggs once more,

8. Nest of the Norwegian wasp (Vespula norwegica) in a black-currant bush.

9. Nest of the same species opened to show structure.

10. Part of a fence from which wasps have rasped away wood-fibre for nest-building ("paper-making").

11. Comb from the nest of a wasp (Vespula) showing closed pupa cells and other cells open.

12. Queen German wasp (Vespula germanica) found hibernating.

13. Polistes wasp feeding larva. Two other larvae are clearly seen and some unhatched eggs. This wasp and those in 14, 15, 16 and 17 belong to an American species of the genus Polistes.

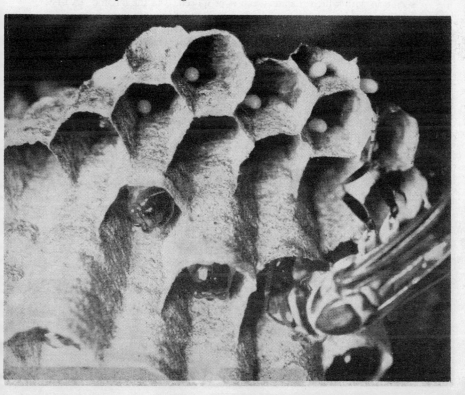

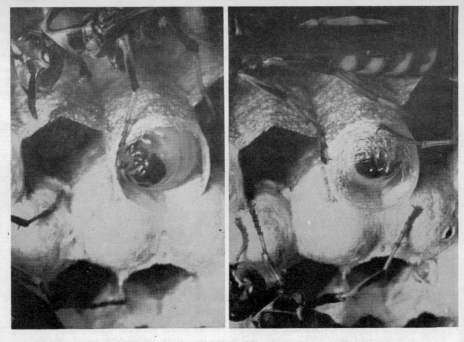

14 and 15. When a wasp larva is fully grown, its lower lip exudes a sticky secretion with which it spins a cap over its cell. On the left the larva has begun spinning and on the right the cap is half done.

16. Polistes worker in flight by nest. The photographer reports that this wasp had waged " successful warfare " against him.

surrounded by a ring of cocoons. After a few cycles of this sort the arrangement tends to get out of step, and any comb in which all stages are found mixed together is known to be an old one.

The development from egg to adult takes about thirty days ; the exact time depends partly on the temperature and partly on the age of the nest. A comb which has the eggs, grubs and cocoons irregularly arranged must be at least ten weeks old. A cell which has been used before can always be recognised under the microscope by the remains of the thin silken lining which formed part of the cocoon, and by the dried larval excreta which accumulate at the bottom of the cell.

A successful nest in the late summer may contain thousands of cells and perhaps two or three thousand workers. A nest of the common wasp, which I dug up in a field at Slough on September 11, 1935, contained a queen and 3,008 workers. It consisted of eight combs, with 11,299 cells in all. The fifth and sixth combs were the largest, each with about 2,188 cells. Nearly all the cells contained brood in various stages, so that many more wasps, including young queens and males, would have been produced later.

No nest with more than one or two hundred workers can be safely examined without destroying the wasps. This can be done most conveniently with the dangerous poison potassium cyanide : a piece as big as a walnut placed at the entrance will rapidly kill every adult wasp. Carbon bisulphide and some other poisons are almost equally effective. The Slough nest was not unusually large, and much bigger ones, with up to ten thousand workers, are found in some years. In such nests a great deal of food has to be brought in, and large numbers of flies and other harmful insects are consequently destroyed. Unfortunately, wasps are not very particular about what they catch, and they will pick the blue butterflies off their evening resting place on grass-stems, besides other insects, or butcher's meat if they can get it. Wasps need sweet stuffs and water. It is doubtful if they often damage sound fruit, but they will enlarge any hole started by other insects or by birds. Soft, over-ripe fruit is readily attacked and wasps are very fond of blackberries in the hedges. They visit flowers not only to catch flies but also to get nectar, being very fond of figwort and wild parsnip.

Wasps have a sting with microscopic barbs, but these do not prevent them from withdrawing it after use. They use the sting only in defence, especially of the nest, rarely for killing their prey—an interesting social trait distinguishing them from the solitary species. One soon finds that a wasp at or near its nest is a much fiercer enemy than one flying about in the open. Association with the nest seems to call out a characteristic aggressive behaviour pattern. Moreover, the wasps from larger nests are fiercer, but this may be because such nests are usually hotter inside and this increases all aspects of insect activity.

THE WASP SOCIETY

Several entomologists have measured the temperatures of wasp-colonies. According to Himmer, a thriving colony of the common wasp runs at 26–36° C which is 5–15° C above the temperature of its surroundings ; the temperature varies much less than that of the air outside. The protected situation of most nests, and the good insulation provided by the envelopes, make some degree of temperature control easy, since any heat produced will be lost only slowly. The numerous larvae continually digesting rich food and the crowds of active workers are of course at all times providing the necessary central heating.

Since the length of the developmental period, and probably also the egg-laying rate of the queen, depend on temperature, the control of temperature is very important. It is well known that the developmental period is shorter in large colonies than in small ones. If the nest gets too hot some of the wasps stand at the entrance and fan with their wings. This has been shown under experimental conditions to be moderately effective in the hornet. When a nest was heated with an electric heater the wasps began actively fanning at 37° C. This kept the temperature steady for a few minutes. When, in spite of fanning, it began to get hotter, they flew off and abandoned their brood.

We may now consider the behaviour of the colony in two of its most distinctively social aspects, the relation of the workers to the grubs and the division of labour amongst the workers. The larger wasp grubs are able to call attention to themselves by bending their heads backwards and scraping the wall of the cell with their mandibles. When a number of them do this together quite a distinct rustling sound may be heard. It is possible that they do this more often when they are hungry. But they have another and probably more important way of attracting the workers. The labial gland, which when they are full-grown produces silk for the cocoon, produces earlier in life a sweetish secretion. This appears as a little drop on the lower lip, and can be elicited by stroking the grub's head with a straw. The workers lick up the secretion whenever they feed the larvae. It has been suggested that this mutual exchange of food is a very important factor in maintaining social ties, but there is very little evidence that the liquid obtained from the grubs is of much consequence to the workers. It may be that the process is really of more importance in nest-hygiene, the workers removing liquid that the grubs could not otherwise dispose of conveniently.

It is clear that in a large flourishing wasps' nest there are many different jobs to be done, and one would like to know whether there is any division of labour amongst the workers. In one sense there evidently is such division, since at any one moment the workers are not all doing the same thing. Some are standing guard over the nest, others are feeding the young, others getting food or paper. The observations of Weyrauch in Germany suggest that an individual may tend to do a particular sort of work for some time, perhaps a day or two, but that there is no rigid division of labour. If some particular task, such as defending the nest, becomes temporarily more important, all will take part in it, and in any case during its life-time one worker will probably do a little of everything. In the honey bee, in which the workers are all very similar to one another in structure, there is a division of labour largely based on the age of the individual, each worker tending to perform different duties as it gets older. In the wasps, the workers are also very similar to one another, but the relation between age and duties has hardly been studied. According to Gaul, the younger wasps work in the nest and do not go out to forage, but from Weyrauch's records of the relatively rapid changes of occupation, it is not likely that there is an elaborate cycle of duties.

MALES, QUEENS AND WORKERS

Towards the end of the life of the colony the workers suddenly start building larger cells. The line between these and the smaller ones may cut sharply across the bottom comb, but any combs added later below this will consist of large cells only. The young queens come exclusively from eggs which have been laid in the large cells. The males develop in either sort of cell, but only in the lowest of the small ones. How the queen comes to distribute her eggs in this way is not known. The whole success of the colony is measured by the number of large cells and therefore of young queens which it can produce. In a temperate climate with a limited breeding season, there must always be an optimum arrangement. If worker production in small cells goes on too long, there may not be time before the frosts to rear many queens. On the other hand, the larger the nest and the greater the number of workers, the more queens can be reared.

,Weyrauch has published records of the number of queens and males produced by colonies of the different species of wasp in Western Germany. The figures show that the bigger the nest and the more combs of worker cells, the more queens are produced. On the average, something between twelve and forty worker cells correspond to each queen cell, and these figures give some measure of the effort required to produce a large crop of queens. The colonies produced between thirty and two hundred queens each, depending on the species. The males appear to be about twice as numerous as the queens.

The sudden change in the type of cell constructed corresponds to an important change in the whole behaviour of the colony. After the change, it is supposed, either special queen-producing eggs are laid, or else from then on the grubs are given some special or more abundant food. But the question of what makes queens develop instead of workers is the major unsolved problem in the biology of wasps. There is really no reliable information on which to found a theory. There is no evidence that a special or more abundant food is supplied to the queen larvae. With progressive provisioning, a larva will be given food as long as it accepts it, so that a longer developmental period would mean that more food was supplied and a larger wasp would be produced. But this would only put the problem one step back, since we should want to know why the grubs in the large cells took longer to mature. Janet's records for the hornet show that the developmental period is longer in the early part of the season, but there is nothing to show that queens develop more slowly than the workers which are being reared at the same time.

In the honey bee the queen, although larger than the worker, develops more rapidly, probably because of the richer food she is given as a grub. On the whole, the evidence in wasps suggests that the important thing is likely to be some change in the behaviour or physiology of the old queen rather than a change in the food supplied to the grubs. Common wasps and hornets are the only social wasps in which the queen-cells are easily recognisable; in the others cell size is almost uniform.

When the young queens emerge from their cocoons they mature in a few days and leave the nest rather rapidly. Pairing takes place in the open, probably as often with males from other colonies as with their brothers. Very soon after pairing the young queens enter hibernation. Meanwhile, the social organisation of the colony degenerates. At this stage the workers may kill many of the remaining grubs, either flying out with them and throwing them away, or using them as food for the others, or even eating them themselves. No convincing explanation of this behaviour can at present be given, but from Deleurance's experiments on Polistes it may turn out to be associated with a glandular degeneration of the old workers. The failure of the nurses seems sometimes to be associated with the exhaustion of the ovaries of the old queen or with her death, but it may also occur in nests where she survives. The final end of the colony is usually determined by the onset of cold weather, a few sleepy stragglers being found in the nest up to some date in October. According to Duncan, nests sometimes survive much longer in the mild winters of California, and may even survive the winter.

Besides the sort of cannibalism described above, there appears also to be some eating of eggs, probably both by the queen and by workers. Such behaviour is common in most of the social insects, and a queen will sometimes eat her own eggs soon after laying them. As will be explained in chapter 7, egg-eating may sometimes tend to protect the colony against parasites which try to lay their own eggs in the cells ; some egg-eating may be the price which has to be paid for a type of instinctive behaviour which is at other times socially valuable. It is also certain that the regulation of the number of eggs is essential to the well-being of the colony, and egg-eating might be regarded as a form of birth-control. It is unlikely, however, that either of these explanations can cover all the facts. In most insects some of the B vitamins are essential for growth and normal health, and it is possible that the eggs are a source of such substances. The needs of wasps for vitamins have not yet been studied.

Later in the season the workers, or some of them, may lay a number of unfertilised eggs, and unlike the queen they may lay more than one in the same cell. If the queen dies, worker laying may occur on a much larger scale and will lead to the production of many male wasps.

It is a familiar fact that some years are " wasp years " whereas in other years wasps are scarce. This seems to be due much more to the number of nests which are successfully established in the spring than to the number of queens produced in the autumn. In other words, the hazards of hibernation are less than those encountered during colony-foundation in the spring. For this reason, the campaigns which local authorities in Britain have at times subsidised (e.g. in Berkshire in 1901), paying school-children to collect queen wasps, do not seem to be successful. In 1950, up to the June 15, the Department of Agriculture for Cyprus destroyed 243,394 queens of the local species of hornet, paying rather more than a farthing for each one brought in. Considerable numbers were also destroyed by other interested parties. Later in the same year the Department's officers destroyed 27,829 nests. This was one of the worst wasp years recorded!

Beirne, by comparing the occurrence of wasp years in Britain with the meteorological record, has shown that an important factor is the weather in April, May and June, when the queens are establishing their first nests. Rain at that time is very bad for wasps. Early warm weather followed by a cold or wet spell is also harmful, since the queens all come out of hibernation but are then unable to establish themselves. Prolonged cold weather which merely prevents them from leaving winter quarters is harmless. Nixon has maintained that, apart from weather, the most important enemy of wasps is other wasps. Suitable nesting-sites are usually not sufficiently numerous and attempts at usurpation lead to fights in which one or both queens may be killed. It is quite common to find a dead queen lying in the tunnel leading down to the nest. Other enemies are badgers which dig up nests to eat the grubs : evidently their fur protects them from stings. It is doubtful if this enemy is common enough to be a serious check on numbers.

WASPS IN BRITAIN

There are five other species in Great Britain, all superficially very like the two common wasps. One of them is a parasitic wasp without workers and will be mentioned in Chapter 7. Another, the Norwegian wasp, builds nests suspended in bushes or trees instead of underground. It is considerably commoner in the north and west than it is in south-east England. The hornet, Vespa crabro is much larger than the other species, and has lately been getting much more common. It is chiefly found in southern, especially south-east England, and twenty years ago was rarely seen, except in the New Forest and in some parks near London, such as Richmond Park. It is now quite common all round London, though being so conspicuous it does not take many to produce an impression of numbers.

It seems that the hornet is not nearly so ferocious as its appearance suggests. It is much the largest of our wasps, and its colour makes it very conspicuous; this may indeed have some value in preventing attacks by insectivorous birds. But from the human point of view what matters is the willingness of the wasp to use its sting. I have never been stung by a hornet, and I have been told by a friend who had the task of removing a nest from the roof of a thatched cottage that he was able to do this without wearing any protective clothing and without being stung by the workers which were flying around. One certainly cannot do this with the larger nests of the common wasp, and this may be partly due to the relative numbers of workers. As we have seen, the more workers there are, the more aggressive their behaviour, and a hornets' nest rarely has more than a few hundred inhabitants.

Wasps very like the British ones are found in all countries with temperate climates, except in South America and the Australian region. The tropics are mostly occupied by social wasps of other types. The hornet, though itself an inhabitant of temperate climates, belongs to a tropical group found chiefly in Asia and the East Indian islands. The hornet sometimes works very late on warm nights. Amongst its Malayan allies are certain very peculiar wasps with a pale ghost-like colour and enlarged eyes which are entirely nocturnal. Very little is known as yet of the biology of any of these tropical hornets. It may prove to be interesting, since there is already a hint that in them the queen and worker are much less different from one another than they are in the temperate species.

3

VARIETIES OF WASP SOCIETY

The social wasps described in the previous chapter are familiar because they are common in those parts of the world in which most naturalists live. In other parts of the world, especially in the tropics, several other groups are found. Two of these groups, the Polistes wasps and the Polybiine, wasps, are of special interest because of the light they throw on the nature of social life and on its probable mode of origin in the family.

THE POLISTES WASPS

The Polistes wasps are found almost everywhere, both in temperate and in tropical climates. Though common in North America and in Western Europe, they do not occur in Britain except as an accidental introduction. They have been much studied, since the whole history of the colony can be followed with ease. This is because the nest is small, is suspended above the ground, and is not surrounded by an envelope. The European species build a nest rather like a single, irregular wasp-comb, usually of less than one hundred cells, but occasionally in the south with up to five hundred.

In temperate climates the life-history is like that of the common wasp. In external appearance the workers can hardly be distinguished from the queen except by their behaviour. It is, however, possible to mark each individual on a comb with different coloured paints so that each one can be recognised. It is then found that one individual, the queen, stays in the nest and does all or almost all the egg-laying. She receives a relatively large share of the food brought in by the foragers. The workers do all the food and paper gathering and most of the nursing. The workers are rather smaller than the queen, on the average, but there is a wide overlap in size. If the wasps are dissected, the queen is found to have a much better-developed ovary and to have sperm in the receptacle where they are stored. Neither of these features, however, may be present in the young autumn queens before hibernation, and these can be distinguished from the workers only by their inactivity. The workers' ovaries are smaller than those of the old queen though better developed than in the workers of the common wasp. Workers are never fertilised. No special queen cells are made when the colony is mature.

Deleurance, who has kept nests in his laboratory in Paris, has shown that if the queen is removed the workers may begin laying male-producing eggs within twenty-four hours. This proves that their failure to do so in the ordinary way is not because they are unable to lay eggs, but because their egg-laying is inhibited by the presence of the queen. This inhibition may be partly due to active interference by the queen. To lay an egg a wasp has to enter a cell backwards, and this action might stimulate the queen to threaten or to attack. In nature the queen is rarely observed to make such attacks, and probably her mere presence inhibits egg-laying by the workers. This interpretation is supported by Pardi's observations in Italy. In many nests the queen dies rather early in the history of the colony, and the workers then begin laying eggs. Pardi found that in late summer some nests produce queens and males in about equal numbers, whereas others produce nearly all males. The second sort of colony is one in which the queen has died early.

At the end of the season many of the grubs in the Polistes colony die and may be thrown away by the workers. This seems to be much the same as the autumnal destruction of grubs by the common wasp, though in Polistes they are not killed by the workers. Deleurance has obtained good evidence that the production of "aborting brood" is due to some lack in the food provided by the workers, probably in some glandular secretion which they are no longer able to supply. By rearing his wasps in captivity at a constant temperature, he was able to have colonies at all stages of development simultaneously. He could then show that brood from young nests aborted if given to old wasps and, conversely, brood from old nests could complete its development successfully if the old workers were replaced by young ones. Since, however, in both experiments the wasps were given plenty of insect food, abortion cannot be due to simple scarcity. It is probably due to the exhaustion of one of the glands whose secretion is mixed with the food. This idea is supported by an experiment in which the workers were kept cool (4° C) at night, so that they did not age so rapidly. With this treatment, brood could be reared normally in the laboratory for seventy days instead of the usual maximum of thirty-five days.

The species Polistes gallicus is particularly interesting because its habits have been studied over a wide range of climates, extending from north Germany to the oases of the Sahara. In the north, colonies are usually founded in the ordinary way by a single queen after hibernation. In south Germany, more than one queen is often seen on a nest. Heldmann noticed that several queens might begin to build nests on one beam of a shed. When none had more than three or four cells, there might be a desertion by one queen who would join another, so that some nests were eventually founded by two or three queens. At first all such queens took a nearly equal share in the work and each of them would lay eggs. But after a time one queen became dominant and did more and more of the egg-laying, leaving the nest less and less often. The others gradually ceased to lay eggs and took on all the foraging duties.

In north Italy, Pardi found that nearly every colony of this wasp was founded by three or four queens. He has described a process of active "terrorisation" by which one of them becomes the real queen. She will attack the others with open jaws, making a loud buzz. After a time she establishes her position so securely that force is not required, and the ovaries of the other queens degenerate, either because the wasps work harder and have less food or because they are not allowed to lay. Amongst these worker-like queens, which Pardi calls auxiliaries, there is a regular order of dominance, in which the queen always comes first. The position of a wasp in the scale is shown most clearly in the disposition of food, but it appears also in the frequency with which it shows aggressive behaviour and lays eggs, and in the amount of time it spends on the nest. A wasp high in the scale is more likely to be given food by one of its inferiors and less likely to give it away. Similar orders of precedence are known in many mammals and in birds, such as chickens, which live in groups. According to Pardi, when the first brood of workers emerges, the auxiliaries all disappear, so that they serve only to get the colony started and as far as is known never found a colony of their own.

In the Sahara, Weyrauch found that this same species of Polistes founded colonies by swarming; that is, the founder queen is accompanied by a number of workers. This is possible only in climates where breeding is more or less continuous. In the tropics also, many species of Polistes seem to swarm, though colonies founded by single queens are common too. It seems probable that in Polistes swarms the queen is a young, recently fertilised female and not the old mother as in the honey bee. Much more study of this point is still required.

This variation in the behaviour of Polistes from north to south may provide an important clue to the origin of social behaviour. A colony in which only one female lays eggs loses a large fraction of the potential birth-rate; the compensating advantage is the greater safety and better care of the young. Any auxiliary or worker is an egg-layer lost but a nurse or hunter gained. According to what is the most likely reason for failure of the wasp-colony, it may pay to sacrifice more or fewer queens in favour of nurses. In the north, where climate is perhaps the chief difficulty to be met, direct multiplication may be more important. As climate becomes less severe, it is possible that enemies of various types become more numerous, and it becomes advantageous to sacrifice up to two-thirds of all the queens in order that nests shall be safely founded.

The process of swarming depends in the first place on a climate sufficiently favourable to allow continuous breeding, or on a worker caste capable of hibernating like the female. On the southern edge of the South American tropics, where the winter is short, whole swarms of Polistes apparently do hibernate together. (The honey bee, of course, withstands the winter by storing honey for the period when food cannot be obtained.) A few of the Polybia wasps described below are able to do this too, but Polistes never store more than small drops of nectar in some of the cells. Swarming involves a great sacrifice of direct multiplication, since every worker which accompanies the queen has been bred from a cell which might otherwise have produced another queen. One hundred cells, at the end of the season, might produce ten queens, ten males and eighty workers. The males would fertilise the females and die, and ten swarms of one queen and eight workers could be produced. On the other hand, in the absence of swarming, something like fifty males and queens might have been produced, with the potentiality of six times as many new colonies. The only possible conclusion seems to be that swarming is advantageous because the colony is at all times protected by a number of workers.

Owing to its small size, to the absence of an envelope, and to the relatively exposed situation in which it is built, the colony of Polistes cannot avoid being cooled at night. In Europe, these wasps usually build the nest in a sheltered, south-facing situation, so as to get the full heat of the sun, and they have two effective methods of cooling it if it gets too hot, as it sometimes does when the sun's rays fall directly on it. In the first place they fan very vigorously with their wings. If the temperature continues high, some of them fetch drops of water which they put into the cells. The fanning is continued and the nest is cooled by evaporation. The nest may be maintained at 12° C below the temperature of an unoccupied nest placed nearby for comparison.

According to Deleurance, the cooling of the nest at night plays an important part in determining whether a grub will turn into a worker or a young queen. He had earlier shown that in artificial conditions queen-producing eggs might be laid and the grubs might even have spun their cocoons before any workers had hatched, so that the developmental stage of the whole colony does not seem to decide whether or not

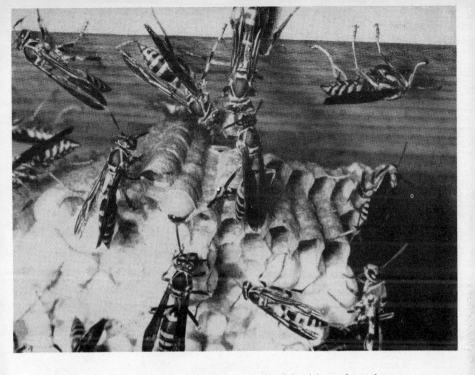

17. A group of three Polistes wasps share a load of food brought in by a forager. Other wasps are feeding larvae or adding to the comb.

18. A Polistes wasp and its suspended nest. The nest has no outer cover like those of more advanced wasps.

19. A hornets' nest. Contrary to proverbial lore, hornets are not especially dangerous; but they are larger than wasps, and correspondingly more alarming.

20. A hornet worker (Vespa crabro).

21. Another example of Polistes wasps and their nest, taken by the author.

22. Nests of Polybia rejecta, hanging from a mango tree.

23. Close view of nest of Polybia rejecta.

24. Nest of another species of Polybia (P. occidentalis).

"queen-eggs" can be laid. Later he kept a number of nests warm by day, but exposed either the grubs, or the adult wasps, or both, to hot ($25°$ C) or cold ($5°$ C) conditions at night. Thus there were four experimental conditions : grubs hot, wasps hot ; grubs hot, wasps cold ; grubs cold, wasps hot ; grubs cold, wasps cold. He found that when the wasps were kept cold at night, the grubs turned into workers, but when they were kept warm, into queens. The temperature at which the grubs themselves were kept was immaterial. This suggests that temperature in some way alters the behaviour or physiology of the workers so that they give the grubs different food. Nobody has yet detected any special queen-food in wasps, but a change in the type of secretions produced by the workers and mixed with the ordinary food might be very difficult to demonstrate. It is not yet possible to relate these experiments in any simple way to what happens under completely natural conditions.

THE POLYBIA WASPS

Most wasp biology was studied first in Europe and much more is known about the behaviour of the species of cooler climates than of those in the tropics. This has led, almost by accident, to the view, held at least implicitly, that the normal thing is to have an annual colony, started by a single queen. The truth is, however, that the majority of wasps, as of most other kinds of insects, live in the tropics, and it is much more likely that social life arose there. When it is realised that social life, at any rate as soon as it makes any real forward step, involves the sterilisation of a considerable proportion of potential egg-layers, it seems almost certain that these organisations first arose in warmer climates. Except in conditions which make continuous breeding possible, many of the advantages of a social organisation are lost. The kernel of the argument is that the efficiency of this type of organisation, measured by its rate of reproduction, increases rapidly with time. In general, short-lived colonies are less efficient than ones which last longer. Thus the winter rest which is imposed by the temperate climates raises a very real difficulty. It seems likely that this has been overcome in a few species which have managed to evolve special modifications. The relatively large size of the queen in Vespula may be one of these; so also is their ability to lay up a large store of internal fat which keeps them alive during hibernation.

The probability that social wasps first arose in the tropics is one reason why it is interesting to examine the Polybia wasps, the largest tropical group of social species. Most of them are found in Central and South America, but a few also in tropical Africa and Asia. One or two penetrate a little way into the United States. In 1937, my wife and I made a special study of them in British Guiana, and much of what is written below is based on personal observations. The general picture of their biology was already known, especially from the work of the von Iherings and of Ducke, but there were a great many gaps to fill in matters of detail.

With few exceptions these wasps found their colonies by swarming. The swarms seem to consist of young fertilised queens accompanied by a large group of workers. Since males are never found in young nests, it is probable that they die after fertilising the queens. Usually several egg-laying queens are found in each nest; the number varies a good deal even in a single species and some colonies may have only one queen. More often there are four to ten, and in some species a hundred or more. It is hardly ever possible to tell the workers from the queens at sight without dissecting them to examine the ovaries. In some species, however, the abdomen of the older queens becomes swollen after egg-laying; they can then easily be identified. In other species, especially in the ones where queens are numerous, most of the wasps have ovaries more or less developed and it may be possible to decide which are queens only by seeing which ones have been fertilised.

Most of these wasps build nests hanging from or attached to trees, and they are often nests of very beautiful construction. Sometimes the outer envelope is covered with lichen or bits of moss so as to make it almost invisible. There is also wide variation in the plan of the nest. In the commonest type each new comb is built on the outside of the lowest part of the envelope and is enclosed by building out from the envelope all round its edge. The exit hole, which is at the bottom of the envelope, pierces each successive comb. All round their periphery the combs are fused with the envelope, and the only way the wasps can get from one to another is through the hole which was the exit hole at an earlier stage in construction.

According to the species, there are probably two sorts of colony, short-lived and long-lived. The former break up into new swarms after a few months' growth, whereas the others last for several years. It has been claimed that in one Brazilian species colonies may persist for twenty-five years, and records of three or four years are certainly authentic. The difference between these two types of colony depends on the physiology and in particular on the length of life of the queens, but it may also be related to different ways of attaining efficiency in reproduction.

We were very impressed by the dangerous life which these wasps lead, especially by their constant exposure to attacks from ants. Not only are they surrounded by a very much greater variety and number of ants than one ever sees in Europe, but the hordes of driver ants, which live exclusively by hunting, are peculiar to the same parts of the world as the

Polybias. This danger can be met in several ways; perhaps the best is to have friendly relations with the ants. A few species have achieved this, in what way we do not yet understand, but they build their nests almost exclusively on trees harbouring enormous ants' nests. These ants do not molest them and the wasp colonies are in many cases long-lived. Several kinds of birds have learnt to build their nests on the same trees and seem to have the same immunity. Probably the small, short-lived type of colony is an illustration of another way of meeting the same difficulty. A small nest does not represent a large stock of capital and when the ants make an attack the wasps can desert the nest without irreparable loss. The queens, because there are several of them and none of them has been laying at a very high rate for long, have lost none of their mobility. They are not nest-bound like the old queen of the common wasp. With the aid of the swarm of workers, a new nest can be reconstructed in twenty-four hours. We witnessed an actual example of such an escape near the Kaieteur Falls. A few days after we had found a nest, we noticed one morning that it was deserted except for a few workers which were biting away the envelope. Tracking these wasps down, we found a new nest about ten yards away (Plates 25, 26, facing p. 80). The cells of the old nest were perfectly clean inside and we assumed that it had been raided by driver ants during the night.

The advantage of the Polybia method of colony foundation could be seen very clearly. The swarm consisted of twenty-two queens and 118 workers and in two days they were able to build a new nest of seven combs with 167 cells, each one of which contained an egg. At this point, the nest was raided again by driver ants, the tail of whose retreating column we witnessed, and the cells were cleaned out once more. This time we caught all the wasps for dissection.

In contrast to the chequered history of this colony is the group of nests of Polybia rejecta which we found hanging from a mango-tree. This tree was covered with enormous ants' nests, probably the separate houses of a single gigantic ant-city. The ants had built a nest of carton, a material composed of earth and wood fibre which is formed into chambers hanging from the branches of the tree. The ants swarmed everywhere, so that it was extremely unpleasant to climb on the tree, though the ants, a species of Dolichoderus, do not sting, but only bite and squirt poison. On the same tree were at least eight large nests of the Polybia, some of which were certainly one or two years old and probably a good deal older. From the outer branches of the tree were hanging several nests of the Oriole bird or Cassique which frequently joins in this association. It was clear that the Dolichoderus for some reason did not attack the wasps, but they would certainly defend their tree if any horde of driver ants tried to invade it. It is thought that the ants are the first to colonise such trees, and that they provide most of the mutual protection. The wasps come later, and though their stings may make the tree unpleasant to some possible enemies, such as monkeys, they seem to be almost helpless against ants.

The size of the Polybia colonies is as variable as their appearance. One large group of species make small colonies like those of Polistes, with only a few dozen cells and often only half a dozen wasps. It used to be thought that these colonies were founded by a single queen unaided. But we found that though this was often true, sometimes a very young nest would be occupied by two wasps. Since the cells clearly had never contained a cocoon, neither of the wasps could have been bred in it, and there must have been two founders. We also found that some nests had more than one egg-laying queen, so that the real difference between these colonies and the more usual Polybia type was in their small size rather than in the number of queens or in the absence of swarming. The normal nest of a Polybia wasp in British Guiana consists of 300–3,000 cells with 50–1,000 wasps. One of the moderate-sized nests of Polybia rejecta consisted of 7,758 cells and contained about 1,400 wasps. The most populous nest we studied, belonging to another species, had about 21,600 cells and 7,087 wasps of which 93 were egg-laying queens. It seems to be a rule that Polybia colonies are larger in the sub-tropical regions than they are in the rain-forest near the equator. Some, for instance near Rio de Janeiro, may be very large, with probably tens of thousands of wasps. Some of the nests are over a yard long and hang from trees like gigantic vegetable marrows.

Most species of Polybia seem to feed on other insects. Sometimes one finds a nest which is temporarily packed tight with winged ants or winged termites. These are caught on their nuptial flight and are probably stored; such flights are not very frequent and the opportunity is taken to catch a large number. Another group are active hunters for insects and are found foraging widely all over the forest. These are bad stingers and defend their nests with great ferocity. They also bring home pieces of fresh meat or of carrion if they can find it. Yet other species are flower-visitors and collect nectar or fruit juices; probably most Polybias do this to some extent. A few have become regular collectors of nectar, like the honey bee. The best known species with this habit has been more or less domesticated in Mexico. The colonies often hang in a citrus bush; they are allowed to grow to a certain size, and are then cut off or smoked out, and the honey is drained off. If a stump of the old nest is left in the bush the wasps will rebuild in the same spot. Apparently the same kind of wasp in more tropical climates stores very much less honey. The store of honey may well have one of the same uses as it does in the honey bee, namely that of allowing the wasps to get through a period (either dry or cold) when food is scarce.

It was stated earlier that Polybia workers could as a rule hardly be separated from the queens. A curious fact which we discovered was that in several species careful measurement showed that the workers and queens differed somewhat in their proportions. The queens are a little larger and the base of the abdomen is often relatively broader. These differences could not arise after the wasps are adult, so that it seems that, as in Polistes and the common wasp, the difference between queens and workers is determined earlier in life. There may well be some difference in feeding, though there is as yet no evidence for it. We found that the number of workers available to look after each grub increased with the age of the colony. This is a simple consequence of the fact that a worker lives longer than the time it takes for an egg to turn into an adult, so that the number of workers relative to grubs gradually increases. It may be that the very slight differences between queens and workers are produced by the increased care which the grubs receive during the last part of the life of the colony. This would be the simplest way of differentiating queen and worker castes and would be automatic, without any demand for special behaviour.

It seems clear that there must be some balance between egg-laying and the construction of cells since either the queens would lay below the rate of which they were capable or else many cells would be built for which eggs could not be provided. We found that in Polybia nests on the average about one-fifth of the cells at any time are empty. It was possible to show that the rate of egg-production is probably the limiting factor in the growth of the colony. The age of the smaller colonies can be roughly determined by examining the contents of the combs and by seeing how many cells are being used for a second time. If the age of the colony is divided into the number of cells it gives a measure of the rate of construction—so many cells per day. This rate depends partly on the number of workers in the nest and partly on the number of queens and on the total of eggs they have produced. But of course the number of workers themselves, so far as they are not members of the original founder swarm, also depends on the number of eggs which has been laid. It is, however, possible by a statistical analysis to show that the number of queens, and therefore the egg-supply, controls both the growth of the colony and the eventual number of workers. What is not yet clear is how a large number of queens together limit their egg production. It may be that production is controlled automatically, from the necessity to share some essential food which is limited in quantity.

WASPS IN GENERAL

It is generally held that the wasps have the least advanced type of social organisation of the four principal social groups (wasps, bees, ants, termites). The worker caste is least sharply differentiated from the queen. The queen is relatively short-lived except in some of the colonies where many queens are present. Foraging still mainly takes the primitive form of hunting for other insects. Such animal food can be stored only for short periods. Thus only in the very few kinds of wasp that store honey can the whole colony withstand a prolonged period of food scarcity.

The behaviour of wasps is rather uniform amongst the different species. Almost without exception they construct hexagonal cells of paper ; this tends to limit the size of the colony, while the collection of wood-fibre involves a great deal of hard work. Yet in spite of these limitations, the behaviour of wasps is sufficiently remarkable. Although there is no advanced type of division of labour, one does find that groups of workers are simultaneously engaged in a number of complicated tasks. This is seen especially in nest-construction, when foraging, envelope-building, and the making of cells and suspensory pillars may all go on at once. It is perhaps easier to understand how the single founder queen may do these things in the appropriate order than how a large number of workers can share out the work at one moment.

The behaviour involved in temperature regulation is particularly noteworthy. When the nest gets too hot, it appears to be only some of the workers which take the steps necessary to cool it. In a more typical example of instinctive behaviour, all individuals exposed to apparently identical stimuli might be expected to do the same thing. We do not yet know whether the workers which go and fan the rest are those which have themselves temporarily got hotter than the others, or whether they take up the work for some other reason. Fetching water to cool the nest, as in Polistes, raises a further problem, since the wasps are not cooling themselves but only the nest they sit on.

The workers are, it is true, very little different from the queens, except in a few species like the common wasp. Even here the difference is much less than in the honey bee or ant. It is clear that we do not yet fully understand what turns a grub into either a queen or a worker. It certainly looks as if it is mainly due to automatic factors, such as the growth of the colony, or of external ones, such as temperature. It is perhaps wise here to make a qualification. There is always a possibility that when two possible types of development, such as those producing a queen or a worker, have been evolved, the actual trigger which sets off the individual on one or other path may change. It is conceivable that even if temperature now controls the development of queen or worker in some Polistes it was something else which had this effect originally.

There are really two problems: first what is the evolutionary advantage of having a worker caste? Second, what mechanism leads to the production of workers at the present day? The advantages are fairly certain, namely a more accurate control of egg-production and the release of energy from that process for the other needs of the colony. The mechanism of the development of the two types is much less certain and need not be the same in every species.

Most species of social wasps live in the tropics, and species like the common wasp or the European Polistes, though extensively studied, are really exceptions. It is probable that the primitive and simpler arrangement is to have a colony with several queens, with workers very little differentiated, with no queen-hibernation but with colonies breaking up into swarms.

The life-cycle of the common wasp, with annual colonies and hibernating queens, seems to be a modification for a temperate climate. One very important feature is the inhibition of the young queen's ovary in the autumn. The quantity of food which the queen takes before the winter is mostly laid down as fat to sustain her during hibernation. In the spring, however, a similar process of feeding leads to egg-development instead. This looks much more like a special device for meeting the unfavourable climate than an original feature of social life. The common wasp seems to be much more firmly attached to temperate conditions than is Polistes, and if they are both regarded as invaders from the tropics, the invasion of the common wasp and its allies must have taken place a much longer time ago.

Finally, the dominance of the queen in a wasp colony is very marked. In her absence the social organisation tends to degenerate. Many workers may start laying and the rule of one egg to the cell may be broken. When the queen is functioning normally she seems somehow to determine the rate of cell-construction, and the length of her life seems to decide how long the colony will last. There are no grounds for thinking that she " gives orders " to the workers, except to the limited extent that she may prevent illegitimate oviposition, but she seems to provide the central point round which the whole complicated pattern of behaviour revolves. Thus, the simple family character of the social group is more fully preserved in wasps than it is in ants, honey bees or termites.

4

SOLITARY AND SOCIAL BEES

Few insects are more familiar than the honey bee, one of the very few insect species which has been domesticated and which is consequently the subject of a large literature. Nevertheless the vast majority of bees are solitary species and rarely noticed except by the specialist. There are two hundred species of bees found in Great Britain, but less than one-quarter of these are social. In most of them, such as the Anthophora and the mason bee shown on Plates 27–30 (facing p. 81), a single female constructs a nest in the soil, in rotten wood, or in some crevice, quite in the manner of a solitary wasp. Sometimes, as in the mason bee, many cells may eventually be constructed side by side. There are good grounds for thinking that in the remote past the bees evolved from the same stem as the sand wasps, the beginning of the divergence being a change in habit from hunting to collecting vegetable food. Many wasps make considerable use of fruit-juices or of the nectar of flowers and one curious group of solitary wasps (Masaridae), found in warmer climates, lives entirely on pollen and nectar in all stages. The bees made a similar change and developed, gradually, a series of new bodily structures which especially fit them for this mode of life. Many of the hairs, which are usually numerous all over the body, are branched. This helps the pollen grains to adhere to them and enables a large load to be carried on each journey. Often the hairs are grouped on special areas of the legs which may be considerably broadened so as to make a pollen-carrying device; in other bees most of the pollen is carried by hairs on the underside of the abdomen. The tongue is nearly always more or less lengthened and is sometimes as long as the body. This enables the bee to reach to the bottom of the nectary of flowers and to extract the sweet liquid. This liquid, the nectar, is stored in the bee's crop during the journey back. Here it is partially digested, so that it becomes honey when stored. Solitary bees put a mixture of pollen and honey in each cell before laying an egg in it. Only social species such as humble bees and honey bees store the two separately.

BEES AND FLOWERS

It is well-known that the visits of bees benefit the flower by transferring pollen (male cells) from one flower to fertilise the female cells of another flower of the same species. The nectar acts essentially as a bait to attract the insects, and the insects pollinate the flower while raiding it. For many flowers, bees are the most important pollinating agents, chiefly because of their methodical behaviour and of their long tongues. As will be described later for clover, a crop may fail to set seed if the right kind of bee is absent.

An insect looking after a nest has to work harder than one which is merely feeding itself, like a butterfly or a fly. Thus bees usually fly in a more or less regular way from one flower to another, mostly keeping to one species on each journey and returning to the nest only when they have a full load. This can be shown sometimes by direct observation of individual bees; or, again, it is often possible to identify the pollen of different species of flower and to analyse the pollen loads brought home by bees, as has been done by V. H. Chambers. In the honey bee, individual workers have been marked with coloured paints, so that their field routine can be recorded.

From the point of view of the bee, there seem to be two different ways of collecting food efficiently. One arrangement, seen in some British solitary bees and in a very large number of species in North America, is to specialise on a single species of flower. This involves synchronising the flight period with the flowering time of the plant and has often led to the development of a special structure, of the tongue for instance, adapted to deal with the particular shape of the flower. As in a human population which depends on a single industry, the arrangement may be extremely effective as long as conditions do not change. In the long run, however, it is very risky.

The alternative is to eschew specialisation and to obtain food from whatever flowers are available at the moment. Perhaps most solitary bees behave in an intermediate way: they visit more than one kind of flower but by no means all, and usually have a favourite one. Social bees, on the other hand, are very rarely at all particular, and visit almost any flower which is providing a good flow of nectar or has plenty of pollen. This is seen in humble bees almost as much as in the honey bee. A large colony can find enough food near the nest-site only if it is willing to take what it can find.

SOLITARY ORIGINS OF SOCIAL SPECIES

Michener has lately suggested that social life in most bees and certainly in Halictus and its allies arose by the association of females, usually but not always sisters. Such females may cooperate to build a nest and although all may at first lay eggs, some may eventually take over this function while others become foragers. In one direction, species are known which are completely solitary, in the other are species which form a relatively large colony with a definite worker caste. Although Halictus is not the direct ancestor of the bumble and honeybees it may well show us the stages by which their behaviour was evolved. While we can scarcely hope to find living ancestors of the higher social bees since their evolution started too long ago we may find significant analogies in Halictus.

Nevertheless, one of the more primitive types, the genus Halictus, does still exist and is represented by a vast number of species all over the world. Thirty-six are found in Britain and a much larger number in the United States. Michener chiefly studied those of Brazil and the United States. For reasons which will appear in a moment the fact that they are social was quite unsuspected until about 1923, and many of the details of their organisation are still little understood. One of the chief difficulties in their study has been the large number of species, many of them very alike, especially in the female sex. Many species will often nest together in one bank of earth, and in any ordinary locality in southern England one may expect to find at least sixteen species. Another trouble is the variety of their behaviour, not only when different species are compared but within a single species either in different parts of Europe or in different years.

Some species are really solitary. Thus in Halictus xanthopus both males and females may be found flying together in the home counties during the spring or early summer. Each female after fertilisation enters winter quarters in a hole in the ground. Early next spring she excavates a nest and raises a brood of males and females. The males die soon after mating, while the young females hibernate. Thus there is one generation a year and it flies early. In the same species, in Germany, both males and females may be found flying in the autumn. Probably here there are two complete generations in the year, the females of the second one surviving the winter to renew the cycle in the following spring. In the more favourable climate of southern Europe the males may not only appear in the autumn but also hibernate and appear again in the spring. By these conditions the same males and females may fly both in the autumn and in the spring. Other species, such as the common British Halictus morio, seem to have more than one generation of males and females during the summer, but almost invariably only the females of the last generation survive the winter. In this species, the seasonal picture is as follows:

Spring	*Summer*	*Autumn*	*Winter*
Female	Male and female	Male and female	Female

DEVELOPMENT OF A WORKER CASTE

Life histories of the type hitherto described are solitary, but in some other species the spring female may survive long enough to be found in the nest at the time when the summer females, her daughters, are emerging. Though the mother and offspring do not in this case co-operate in nest-construction, there is here the first hint of social behaviour, since the presence of the mother in the nest must give her young some protection throughout their development. Another group of species, including in Britain Halictus calceatus and several very similar ones, have a real social organisation. Males and females are found in the late summer and autumn, but only the fertilised females hibernate to become the spring females of the following year. The spring females, either alone or working in small groups (up to three females), construct a nest and produce offspring which are all female. This brood of females may or may not be externally different from the spring-mother, but in any case they are unfertilised; their ovaries develop only to a limited

25. Nest of the wasp Parachartergus when first found at Kaieteur, British Guiana.

26. The same colony rebuilding the envelope and combs on a new site after the first nest had been raided by driver ants.

27. A solitary bee (Anthophora acervorum) on a flower.

28. The solitary mason bee (Osmia rufa), male above, female below.

29. Osmia rufa. Larvae feeding on pollen masses in their mud cells.

30. Osmia rufa. View of a many-celled nest built behind a door lock.

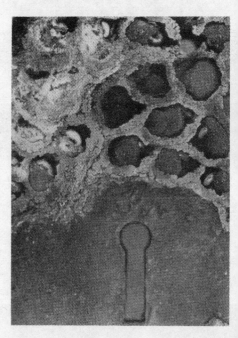

extent, and they are essentially workers. These workers now do nearly all the work of the summer-nest, and the old spring female becomes a queen who guards the nest and lays most of the eggs. The result of this activity is the production of another autumn brood of males and females. It seems from the observations of Stöckhert and Noll that the autumn females all come from eggs laid by the queen, whereas the males come from unfertilised eggs laid either by the queen or by some of the workers. Thus throughout the whole seasonal cycle females are always derived from fertilised eggs which are all laid by the original spring female.

It seems, at any rate in the Halictus studied by Stöckhert, that where two or three spring females combine to make one nest, only one of them survives to act as queen when the workers appear. Thus the situation seems very like that already described for the wasp Polistes gallicus in north Italy, where colonies may be founded by several queens, but in which one queen becomes the dominant egg-layer and is the only one who remains in the nest after the workers have hatched. Apparently in some species, such as Halictus maculatus, rare in Britain, several spring females may construct separate nests with a common entrance-gallery. When such a nest is fully developed, with each separate part containing its own brood of workers, the single entrance is as busy as the entrance of the much larger colonies of a humble bee.

The colonies of Halictus show a type of social life which is primitive because they are always small (probably six to ten workers) while the workers are little different from the queens and still lay some of the eggs. Probably the growth of the colony is much restricted because they have retained the normal subterranean habits of solitary species. Each separate cell has to be excavated, and though a group of cells, looking like a small comb, may be dug up by the entomologist, the comb has not been built up, cell by cell, as in a wasps' nest. Rather the superfluous earth has been removed, and a group of six to twelve cylindrical cells has been isolated in a small underground chamber. The earth is strengthened by a coating of saliva. In Halictus the cells are composed of that earth which the bees have not carried out of the nest, whereas in a social wasp the cells are built of new material which has been brought into the underground chamber. The nearest human analogy is seen when a dug-out is compared with a house.

The great number of species of Halictus, many of which are extremely similar to one another, suggest that the group has evolved recently and is probably still evolving. Their social life is flexible so that even in one species not all colonies have the same make up. It would appear that new types of colony are continually arising from their natural gregarious tendencies.

Rudimentary social behaviour approaching that of Halictus occurs in some other bees of groups which are mostly solitary. Such are the South African species of Allodape which build nests in hollow plant stems. There are no proper cells and the brood lies in a rather disorderly mass with the different stages mixed together. The colony is begun by a single female, but later her daughters help in obtaining food for the brood. So much was established by Dr. H. Brauns, for long a successful observer of African insects. Clearly this example needs much more detailed study before it is fully understood.

HUMBLE BEES

The next social group, the humble bees, are much more familiar and have been much more studied. Apart from the cuckoo humble bees, social parasites which will be mentioned in chapter 5, nineteen species have been seen in Great Britain and thirteen of these are more or less common in the southern part of the country. The group, however, is a very large one, with two hundred to three hundred species, found all over Eurasia to Japan and Sumatra; all over North and South America and in Africa north of the Sahara. Thus, while it is especially well represented in temperate regions, it is also found in much of the tropics. The solitary bees most like humble bees are now found in South America, and it may have been here that the group originally developed.

Some people imagine that a humble bee does not sting like the honey bee. This is unfortunately untrue, but it will very rarely sting unless roughly handled or unless the nest is disturbed. As in the wasps, the larger the nest the more fiercely is it defended. In temperate climates the life-cycle is very like that of the wasp. Males and females are produced in the summer or autumn, but only the fertilised females survive the winter. In the spring the queens feed on flowers of sallow and other early flowering plants, and then begin searching for a nesting-site. Some species nest under grass-tussocks on the surface of the ground, others in holes underground. In both types most nests are constructed in an abandoned mouse or vole nest. The collection of pieces of grass and dead leaves made by the mouse provides the essential protection and insulation for the early stages. More grass or moss may be piled up later.

Suitable sites of this type are often not very numerous, and this has two very important consequences. First, there may be severe competition for nests, and one or even more dead queens may be found at the entrance to the nest, killed by the rightful owner in defence of her property. Secondly, there is a limit to the number of nests which can exist on an area, and this may be very important in the production of such crops as red clover-seed, since humble bees are the chief agents in fertilising the flower. Darwin, in fact, in a well known passage in *The Origin of Species*, records that red clover sets very little seed if the flowers are covered and made inaccessible to bees. He also quotes Newman as believing that mice destroy many humble-bee nests and that near villages and small towns humble bees tend to become more numerous because cats destroy many of the mice. Although this observation needs to be better substantiated, it suggests that there may be a complex, double relationship between mice and bees, the mice being not only enemies but also providers of nesting-places.

The queen, after starting her nest, has to raise her first brood of workers unaided, but once the workers have hatched they do most of the work, especially all the foraging for nectar and pollen. The queen gradually drops all work except the laying of eggs, and eventually becomes incapable of flight. In fact, she never leaves the nest again once the colony is fully established. Sometimes, in bad years, very worn queens are seen flying in midsummer, and these are trying to raise a new brood after the first one has failed to develop. If the nest goes well several groups of workers are raised, and eventually males and young queens. The cycle is then complete and the colony soon breaks up. This may happen surprisingly early, even by late June in some species in a good year. The young June females enter winter-quarters at the height of summer. Apparently there is no time for them to start and to bring safely to completion a new colony in the same season. Bols showed that in Belgium certain banks may be much favoured by the young queens, several dozen digging holes near one another. These hibernation-holes are of about the diameter of a cigar and three inches long. Not all queens use such sites, and some may be found hiding in moss or dead leaves in woods or hedgerows.

Apart from the dead vegetable matter which insulates the nest and helps to protect it from the damp, the cells are constructed of wax, a very characteristic product of social bees. In the humble bee it appears as thin plates protruding between the overlapping segments of the abdomen both above and beneath. The wax plates are removed by a stiff brush on the hind legs, and can be worked up by the jaws. All insects produce at least minute quantities of a wax which serves to make their outer skeleton waterproof. Several kinds, such as scale insects or woolly aphids, produce much wax which seems to serve as a partial protection from their enemies. Only social bees, however, " handle " the wax which they produce.

The humble bee does not construct regular combs of hexagonal cells like the wasp or honey bee, and the nest grows rather piecemeal. This makes its arrangement difficult to understand without careful study. The first step of the queen is to make a small cavity in the centre of the ball of grass, and here she lies for considerable periods while wax is being secreted. As in the honey bee, the production of wax entails the eating of a good deal of nectar. The first wax is used to make a small honey-pot which stands near the entrance to the nest. This when filled provides a food supply during bad weather. The next step is to make a small ball of pollen, moistened with honey. On top of this a small waxen egg-cell is built. In some species the egg-cell is built first and the pollen collected afterwards. The first cell contains six to ten eggs ; from these the grubs hatch in three or four days. Each grub after a time produces a swelling on the sides of the waxen cell, so that the cell is gradually enlarged. In some species the queen opens up the cell at intervals and uses her tongue to inject into it a mixture of honey and pollen. In others the pollen is added in wax-covered lumps at the sides of the original cell. In this way the queen has less contact with her grubs and her behaviour is a modified form of mass-provisioning, whereas the injection of liquid food is a characteristic type of progressive provisioning. In older nests a good deal of wax may be worked into the grass above the cells, to form a waterproof roof.

The grubs are fully fed and begin to spin their cocoons seven to eight days after they have hatched. The cocoons are oval, made of yellowish silk, and in most species stand in a close-packed group side by side. In fact the group of cocoons looks not unlike an irregular comb, though made of silk and not of wax. In the first brood the cocoons on each side are placed higher, so that the central groove in which the queen sits is preserved. Here she spends much of her time keeping them warm. The first workers usually hatch rather more than three weeks after the egg was laid, and the queen or, later in the season, other workers often help them to bite their way out of the cocoon. Those from the middle of the groove develop more quickly than those from the sides.

The new-hatched worker is pale-coloured with somewhat matted hair. Her first activity is to take some honey from the honey-pot. However, in two or three days, she is ready to leave the nest and to begin to collect honey and pollen. Provided enough workers have been raised in the first brood the queen will now stay in the nest until she dies, but some-times, when the first lot of workers is too few, she will con-tinue foraging a little longer.

Later batches of eggs are laid in little pockets attached to the side of the cocoons. Mr. and Mrs. M. V. Brian, working at Glasgow, have recently shown that there is a tendency for the number of eggs in the group to be proportional to the number of cocoons in the clump on which they are laid. Taking the average of cocoon groups of different sizes, there were about three eggs for each cocoon, with six to sixteen eggs in a batch. This seems to be a form of family planning which ensures that there shall be no more brood to look after than there will be nurses to care for it.

At this stage, the brood soon becomes too large to be incubated by the queen, and the groups of larvae or cocoons no longer form a groove but are convex on top. The developing brood produces enough heat from the digestion of its rich food to keep the whole nest at a sufficiently high temperature.

At the beginning of the eighteenth century, a Dutch naturalist, Goedart, first propagated the curious myth of the " trumpeter " humble bee. He observed that a bee often rests on the top of the nest, fanning with its wings and making a loud hum. This often occurred in the early morning when the sun's rays first strike the nest, and Goedart thought that the hum of the vibrating wings was a call to start the morning's work. During the last two hundred and fifty years naturalists have argued whether the " trumpeter " really existed, and if so what were his functions. The Austrian entomologist, Hoffer, finally showed that trumpeting really occurred by observing a colony in a nest-box on his window-sill. Eventually, Plath in America showed that it happened especially when the nest was exposed to the sun, and that this could be checked by moving the nesting-box to different positions. Fanning sometimes occurs also in the evenings and may then serve to dry the nest or to remove odours. Nests in natural situations will less often be exposed to the sun than when they are kept above ground in nesting-boxes, so that trumpeting seems to be observed much more frequently in conditions of semi-domestication. It is behaviour of the same kind as that observed in wasps when their nest gets too hot.

At the height of its activity the colony is able to store considerable quantities of food, especially in some species and especially those nesting below ground in which workers become very numerous. Such nests may have three or four hundred workers, but one hundred and fifty or less is more common. Nectar, later turning to honey, is stored chiefly in empty cocoons which are not used again for brood rearing as are the cells of a wasp-comb. Since the egg-pockets are built near the top of the sides of the cocoons, the nest grows upwards and the empty or honey-containing cocoons are mainly in the lower part of the nest. Sometimes some waxen honey-pots are also built round the edge of the nest, but according to Plath these are temporary structures. The food stored in them is for daily use and they are often broken down and the wax used elsewhere. Some of the species also store considerable quantities of pollen, either in empty cocoons or in large waxen cylinders which may eventually reach the length of more than two inches. The surface nesting " carder bees " store pollen in little wax pockets at the sides of the groups of grubs ; the quantity stored in this way is quite small. Since the colony does not survive the winter there is no point in making a large store of food as the honey bee does. But, as explained later, the store of honey, at any rate, has a very important function in preparing the young queens for their winter rest.

The climax of the colony comes when the males and young

queens are produced. It is not known what exactly determines the date when this shall begin. It is not simply that
the vigorous, successful colony shall have passed a certain
size, for relatively small colonies may produce one or two
queens at about the same date as a larger colony of the same
species produces a much bigger number. It may be that there
is some change in the queen's rate of egg-laying, so that the
number of nurses tends to pile up compared with the number
of grubs to be looked after. Among the brood which hitherto
has produced nothing but workers a certain number of
males appear and a little later queens. After this no more
workers are produced. This sequence, in which males
tend to precede the queens but partly develop among
them, is exactly what is recorded in the common wasp.
The young queens work only in colonies in which the
workers are for some reason scarce. Normally, they stay in
the nest for a time to feed, and then, after pairing, enter
winter quarters.

At this season of the year, one may notice several bees
taking the same course in a garden or wood, though there
may be a few minutes between one flier and the next. At
intervals, they will fly down and hover for a moment in such
places as the hollow at the roots of a tree. These are male
humble bees on their characteristic circuit, usually a different
one for each species. Humble bees have a strong, often honey-
like smell in the male and a less obvious, often rather unpleasant smell in the queen. It has been suggested that the
males emit scent at the stopping places on their circuit and
that these are then attractive to the young queens; the
suggestion has, however, not been studied experimentally.
But R. A. Cumber found that young queens could be
suspended by a piece of cotton in bushes in the circuit, and
that males would then soon come and pair with them. In

some species, however, the queens have been known to pair with a brother, in or near the nest.

In most species queen and worker humble bees differ very little from one another except in size. In the larger, subterranean species the gap in size between the two castes is appreciable and there is almost no overlap. In the carder bees, however, the two types almost run into one another, probably because this group of humble bees has not evolved so far as have the others.

A study of the differences between workers and queens has been made by the author and by R. A. Cumber. A picture of this type of investigation will be obtained if a nest is described which was excavated at Silwood Park, Berks, on July 4, 1951, by Dr. N. Waloff and myself. It was a nest of Bombus hortorum, about six inches underground, in an old mouse-run in a hay field. The bees were divided into three lots : first, those that were out foraging and returned to the nest while it was being excavated, mostly carrying pollen on their hind legs ; second, those that came out to defend the nest while it was being dug out ; third, those that stayed on the comb and were removed from it, one by one, in the laboratory. The weights of these three lots of bees and their status in the nest are shown in the table opposite. Although there is some overlap, the average difference in weight of the three lots of workers is very striking. The old queen (weight 553) was recognisable by her rather battered appearance. Only one of the other queens (weight 442) had been fertilised, as was shown later by dissection. The colony was a rather small one and it may be for this reason that one of the young queens was carrying a load of pollen. Besides these bees, the nest contained six eggs, twenty-four grubs, sixty-six full cocoons, fifty-one vacated cocoons containing honey, and one cell containing pollen. There were also twenty-six empty cocoons not in use. Judging by their size, about thirty of the cocoons might later have produced males or queens. Though there were only forty-three bees in the nest there were seventy-seven vacated cocoons. Most of the difference can be accounted for by the deaths amongst the first broods of workers. This was a moderately successful colony which would probably have finished its effective working life two or three weeks later. Most of the honey would probably have been used in feeding the queens.

Weights of bees in nest of Bombus hortorum, excavated on July 4, 1951. Weights are given in milligrammes (1 mg.=about thirty-five millionths of an ounce).

LOT I "FORAGERS" RETURNING TO NEST			LOT II WORKERS DEFENDING THE NEST		LOT III REMOVED FROM COMB	
Weight of bee	Weight of pollen load	Nature of bee	Workers' Weights	Workers' weights	Queens' weights	Males' weights
395	25	Young Queen	317	202	553	316
332	30	Worker	218	162	493	262
301	45	,,	217	149	465	215
278	62	,,	164	148	462	
268	12	,,	155	140	442	
264	41	,,	150	123	440	
260	0	Male	144	122	432	
220	30	Worker	139	102	398	
155	40	,,	136	96	381	
			115	87	368	
				86		
260 mg.			176 mg.	129 mg.	Average weight of workers	

It seems from this record and from many other similar ones that the variation in size of the workers has some importance in the economy of the nest. The variation can be referred back still earlier to what happens to the grubs. Cumber, by weighing grubs or pupae extracted from different positions in the groups, showed that those in the middle are heavier and those from the periphery lighter. In fact, as Sladen first suggested and Cumber later confirmed, the variation in the size of the workers depends mainly on the competition of the grubs in one pocket for their common supply of food. Grubs which happen to be near the middle of the pocket grow large, while those at each end are smaller.

The variation in size of worker so produced is the basis of the later economy of the nest. It seems that once workers are sufficiently numerous, a division of labour is established between them based on the size-difference which itself depends on the earlier competition for food. Weighing samples of workers throughout the season shows that there is no increase in the *average* weight of the whole set of workers, but when the colony is at its largest, the large bees do most of the foraging; the small ones mostly stay at home to act as nurses. According to very recent work (1952) of Mrs. Brian, small workers do not begin to forage till 15 days after leaving the cocoon, whereas large ones begin at 5 days. Small bees forage mainly for nectar while large bees also collect pollen. Cumber found a further minor division of labour between the foragers, for by measuring the length of the tongue of humble bees caught on different flowers he showed that the largest workers of the species, with the longest tongues, went to flowers with long nectaries, whereas the smaller, shorter-tongued foragers went to flowers in which the nectar was more readily accessible.

In a humble bee such as Bombus agrorum there is a complete overlap in size between the young queens and the largest workers. There is, however, an essential difference in physiology. Cumber found that young queens can be confined in small cages and artificially fed for a week on honey. When this is done they put on weight very rapidly, laying down a large amount of white reserve tissues in their abdomen. Their ovaries, on the other hand, show not the slightest sign of development. Large workers when similarly treated have no power to increase appreciably in weight, and if they are dissected the reserve tissue is found to be discoloured brownish as it is in the old queen.

At least some of the large workers usually show partial development of their ovaries. If there is anything wrong with the queen, many of them lay eggs, though these probably produce males only. We may conclude, therefore, that there is a fundamental difference between the young queens and the workers, caused in some way not yet understood, such that queens can prepare themselves for hibernation whereas workers cannot. There is no evidence that workers ever hibernate in the climate of Western Europe or of North America, and no evidence that they are capable of pairing. These statements have, however, to be rather guarded owing to the difficulty of separating the two castes by mere inspection.

As noted earlier, bees play an important part in the pollination of flowers. This is because they are numerous, are hard and methodical workers, and are relatively large insects, with long tongues. Flowers in which the nectar collects at the bottom of a deep nectary are nearly always fertilised by bees, more rarely by large moths. In temperate climates the bees with longest tongues are some of the humble bees, and it is known that these insects are essential for the setting of seed by such flowers as red clover or monkshood. Both in the Alps and in Norway a special species of humble bee is found which visits monkshood almost exclusively. Without this bee, the flower would hardly ever set any seed.

However, the red clover is a much more important flower to man and has been much more studied. Clovers make a rich fodder-crop and their roots bear nodules in which bacteria multiply and convert nitrogen into nitrates. This means that besides producing good food for cattle, the clover also enriches the soil it grows in. Some clovers, such as the common white species, have relatively short nectaries and can easily be fertilised by the honey bee. The red clover, however, which for some purposes is particularly valuable, has a larger flower and is rarely fertilised by anything other than one of the long-tongued humble bees. This creates a difficulty whenever large fields of clover are grown for seed, since there are rarely enough of the right kind of bee to fertilise all the flowers.

The problem has been studied in Britain at Aberystwyth, in Scandinavia, in Russia, and elsewhere. It was proposed at one time in Russia that wild humble bees should be encouraged by the provision of large numbers of nesting-boxes. For the shortage of bees is partly due to the absence of sufficient nesting-sites, and this may in turn depend on a shortage of mice. For other reasons, too many field mice would be undesirable, so that it was thought that nesting-boxes might be a substitute for abandoned mouse-nests.

Probably the only practical measure is to grow the clover in small patches rather than in the large continuous areas which are much more convenient to the farmer. The problem arose in its most acute form in New Zealand, where there are no native humble bees. For many years the red clover set no seed, but in about 1880 bees were introduced from England. Unfortunately, the introduction was not carefully planned and besides one useful, long-tongued bee, a short-tongued species was brought over at the same time. Some of these short-tongued humble bees have the habit of biting a hole near the bottom of the nectary, so that they can rob it without coming in contact with pollen and fertilising the flower. Once the hole has been made long-tongued species may use it, too, so that much harm may be done. However, for many years New Zealand was able to obtain all the clover seed that was needed, and it is only recently that difficulties have again arisen. It is not clear whether for some reason the long-tongued bee has become less numerous or, as I think is more likely, that the area under clover has been increased without any corresponding increase in the possible nesting-sites.

The humble bees which live in tropical countries have not been much studied, but it has been reported that some of them, as in the tropical social wasps, live in colonies with several queens and found new nests by means of a swarm of workers and queens. This has been recorded by von Ihering in Brazil, but further accounts of the habits of humble bees in South America throw doubts on his interpretation. It seems rather that what von Ihering saw may have been one of the exceptional nests in which the young queens stay in the old nest.

Observations by Plath in the United States support the idea that colonies might become perennial, the old queen being replaced by a new one when she was worn out. In one American species he observed this to take place in two separate colonies at the end of August. In one of them one large crop of new workers was produced in September and a second crop of males and queens in October. This is exceptional behaviour in a temperate climate, and might be possible only in a favourable year.

In New Zealand Cumber found that one of the introduced British species behaved on one occasion in a rather different way. He dug up a large nest on November 4 (which is early in the New Zealand spring) and found that it was a survival from the previous season. The old queen was alive, but was no longer capable of laying eggs, and this function had been taken over by about twenty of the workers. The next also contained more than forty young queens, of which about half had been fertilised, and twenty-one males. In none of the young queens was there any ovarial development, so that they were really only hibernating in the nest. None of them had taken over the functions of the old queen as in Plath's observation. It is an explanation of facts of this sort in terms of physiology which we most need to advance our knowledge of social behaviour. For two of the most important differences between solitary and social species are that first, in the latter, young females stay by their mother, and second, that the development of the ovaries is subject to more elaborate control.

THE STINGLESS BEE

A bee without a sting seems almost a contradiction in terms, but in one large group of social bees, the Meliponidae, the sting is reduced to a vestige and is useless for defence or attack. The insects are, however, by no means helpless, since they can bite and can smear their enemies with sticky substances, so that populous colonies are as much feared as those of wasps. Stingless bees are found only in the tropics and sub-tropics; they are most abundant in South and Central America, but occur also in Africa, Asia, in some of the Pacific islands and in Australia. The species are very numerous : about two hundred are clearly recognised, while recent work in Brazil suggests that there may be many more. The stingless bees fall into two groups. Melipona itself is confined to South America and its species are rather large, often little smaller than the honey bee. Trigona, which is found in the

31. Top to bottom, male, worker and queen of the carder humble bee (Bombus agrorum).

32. Nest of Bombus agrorum, found in a disused mouse nest on the ground.

33. Humble bee (Bombus lucorum) about to take nectar from lupin.

34. Young larvae of a humble bee (Bombus terrestris).

35. Drone honey bee (Apis mellifera). Note the large eyes.

36. A fertile female, or queen honey bee, with large abdomen.

37. An infertile female, or worker honey bee.

Old World as well as in America, has many more species, and varies from a similar size down to some minute bees no larger than house-flies.

Unlike the humble bees, the workers produce wax only from the upper side of the abdomen. Some early workers thought that the males also secreted it, but this has not been observed by any modern naturalist. Besides their own wax, they also collect plant waxes and juices and earth, and these substances are often mixed with the wax for some purposes. The most usual site for a nest is a hollow tree, but some species nest in the open, in the forking of two branches, or in crevices of stonework, or in holes in the ground. A few build in deserted ant-mounds or in holes in large termite-nests. Within such a large group some variety of habit is to be expected, but even within one species more than one type of site may be used. Most of the species construct a spout-like structure at the entrance. This may project for several inches from the hole leading into a hollow tree. It is sometimes closed at night by a temporary wax membrane. In a typical nest, the hollow branch is walled up at each end by a barrier of wax, mixed with earth. The cavity so formed is divided into two, one part for the brood, the other for a store of honey and pollen. The part containing the brood is surrounded by a number of irregular waxen envelopes which doubtless serve, as in the common wasp, to conserve heat. The brood is reared in waxen cells arranged in one of three ways. The commonest is to have a series of horizontal combs, one above the other. The lowest comb is built first, the later ones being supported one above the other on little pillars. The cells all open upwards, being constructed on the upper side of each comb. Occasional cells are missing so that there are gaps in the comb, allowing for the passage of the bees and perhaps for ventilation. In large colonies,

97

several tiers of combs of this type may exist side by side. In another type of nest the combs are arranged like a rather irregular spiral staircase, so that though the combs rise upwards, they are not close-packed one above the other. Finally, in some species of Trigona no regular comb is built at all, but the cells are arranged in a cluster or conical heap. Nests of this type also lack the waxen envelopes surrounding the brood. It has been suggested that bees which made this imperfect type of nest are less highly evolved than the others and have never acquired the habit of building regular combs. This, like many other things about these bees, needs much more study.

Apart from the brood, sometimes on one side of it or sometimes on both, is the store of honey and pollen, but this is not kept in cells similar to brood cells as it is in the honey bee. The storage chamber contains globular or cylindrical pots of honey and pollen. In a large colony there may be a hundred or more of these, each one much larger than a brood-cell. There may also be pots of resin, or plant-materials may be kept in irregular lumps.

The Maya Indians domesticated these bees for their honey long before the Spanish conquest of Mexico. Sections of hollow logs were used as hives, as has also been the custom in some parts of Europe for the honey bee. The stingless bees were introduced into the log by various methods. Sometimes natural wild swarms colonised them. Sometimes the Indians put into the log part of the food store together with some workers and the queen from an older nest. Most commonly, the brood combs of an older nest were divided. Apiaries of a hundred or more logs were kept, and such property was an important part of an Indian's capital and could be the largest item in his will. The Mexicans usually removed honey from their hives twice a year, obtaining about three and a half pints of liquid honey from each. This honey does not seem to thicken and crystallise out even with keeping. Some species or colonies may produce a great deal more : an Australian nest, for instance, has yielded 50 lb.

The castes of the stingless bees are much more strongly differentiated than those of the social species described so far. The worker is the only type structurally adapted for running the colony. For carrying pollen her legs are modified in much the same way as those of humble bees. Her mandibles often bear serrations or teeth which are thought to be of use in manipulating wax. The queen, on the other hand, lacks the pollen-collecting apparatus of the worker, has simpler mandibles and a smaller head with somewhat smaller eyes. She does not produce wax. She would be quite unable to found a new colony unaided, and she must therefore be accompanied by a swarm of workers. In Melipona the queen is no larger than the workers, but in most species of Trigona she is considerably larger; in both groups her abdomen soon becomes very swollen by her enlarged ovaries. An older queen is therefore easily distinguished and is too heavy for flight. In fact she never leaves the nest once it is established.

Many of the details of swarming and of the foundation of new colonies are still obscure. Pairing has not been accurately described, but there is probably a nuptial flight something like that of the honey bee. Probably the founding swarms consist of a single queen and many workers. The laying queen does however tolerate some young queens, and a number of them may be found living in the colony. It is doubtful, however, if there is more than one laying queen at any one time; the others may be merely waiting to take part in swarms.

The way in which the castes come to be differentiated is still a matter of dispute. Many of the species of Trigona build " royal " cells, usually on the edges of the ordinary comb. These cells are considerably larger than the ordinary brood-cells and produce queens which are larger than the workers. In Melipona and in some kinds of Trigona no special royal cells are recognisable and the queen is no larger than the worker. It is also said that she has to feed up for some time before her ovaries can become active.

The stingless bees are almost the only group of social insect in which there is no contact between the adult and the grubs. The cells are filled with a food, mainly honey and pollen, and when the egg has been laid the cell is sealed with wax. There is thus no opportunity for the type of differential feeding which is well-known to occur in the honey bee. Moreover, it is stated that several workers may contribute to the storing of each cell, making it much more difficult to suppose that any accurately balanced diet can be provided. While enlarged royal cells would provide the queen grub inside them with a greater bulk of food, this would not apply to Melipona, since the queens and workers develop in similar cells.

THE THEORY OF HEREDITARY CASTES

Because of these difficulties, Kerr has recently suggested that the characters of the castes are inherited and are not the result of special feeding. When he bred out the brood from combs removed from nests he found that in three species of Melipona the queens formed one-eighth of all the progeny, while in a fourth species they formed one-quarter. These fractions correspond to what are known in the Mendelian scheme of genetics as a " trifactorial " and a " bifactorial " backcross. The meaning of this description can be explained most simply in the species which produces one queen for every three workers, that is, one-quarter queens.

The hereditary factors which are supposed to control the type of development of the individual are represented by letters, the queen having the constitution Aa Bb. During the development of her eggs in the ovary, there is a special cell division (reduction division) in which each group of genetical units (chromosomes) is halved. Thus the Aa Bb queen will produce four types of unfertilised eggs with constitutions AB, Ab, aB, and ab. The males are, as in all bees and wasps, derived from unfertilised eggs, and so have only the half complement of hereditary factors to start with. They omit the reduction division and their sperm are, therefore, of the same four types as the unfertilised eggs. When fertilisation occurs, there will be an equal likelihood for any of the four types of eggs and sperm to meet. Only four out of sixteen possible combinations will give an individual constituted like the queen. Thus sperm AB + egg ab, or sperm Ab + egg aB, would both give a queen. We should therefore expect, on the average, one-quarter of the offspring to be queens and three-quarters workers. Males will be derived from any eggs which are not fertilised.

The case where one-eighth of the females are queens would arise if three factors were involved and the queen had a constitution Aa Bb Cc. While such schemes can theoretically explain the facts, this is very far from saying that they have been proved to be true. Kerr supposes that some similar arrangement may also exist in Trigona and in the honey bee, but that in them the queen grub has to be given the right quantity or quality of food as well as inheriting the right constitution. In Melipona, if queens are produced at a constant rate throughout the life of the colony they would soon become too numerous, and Kerr finds that most of them are killed by the workers.

If Kerr's ideas are substantiated it would seem that

Melipona has a rather wasteful and inefficient method of providing the necessary proportion of new queens. In most groups of social insects the argument whether castes are dependent on hereditary constitution or are due to differential feeding has continued for the last fifty years. In most groups, the weight of the evidence is in favour of control by diet. It will be rather startling if the stingless bees turn out to be an exception, or if our ideas on all the other groups turn out to be wrong.

5

HONEY BEES

The last group of social bees is the most familiar, including
the honey bee and two or three similar species. These have
been domesticated by man from at least the time of the early
Egyptians, and an immense body of tradition and literature
has grown up around them.

It will be convenient to say something first about the
non-European relatives of the honey bee. In India and
elsewhere in the East there are two rather distinct species of
wild bee, Apis dorsata and Apis florea. Our familiar honey
bee is Apis mellifera and itself occurs in a variety of strains
or subspecies. The one which occurs wild and has also been
domesticated in India is Apis indica, while in Africa a variety,
Apis adamsoni, is also often found wild. Apis mellifera must
have been introduced into Europe very early in man's history; it has become markedly different from Apis indica,
though the whole group probably has an eastern origin. The
honey bee is to-day comsopolitan, but this is because it has
been carried everywhere by man during the last few hundred
years.

WILD HONEY BEES

Apis dorsata is considerably larger than the honey bee.
Its colonies build a single comb hanging from the branch of
a tree. This comb hangs vertically and has cells pointing
each way, their long axes being horizontal. All the cells,

including those in which the queens develop, are of the same size. Some of them are used for storing honey, but the colonies are relatively small and do not stay for long in any one site. Apis florea is the smallest species, and it too builds a single comb hanging from a branch. The cells are of four sizes. Near the top are a number of honey cells and below them a band of smaller cells for worker brood. Below this are considerably larger cells in which drones are reared. All these cells are hexagonal, but hanging from the bottom of the comb are some irregularly cylindrical queen cells which are the largest of all.

Apis mellifera in all its forms has a marked tendency to nest in dark cavities, such as caves or hollow trees. Less commonly its combs may be suspended from branches, and this has been recorded occasionally in Europe when a swarm has escaped into the woods. The honey bee builds several combs hanging side by side, each somewhat resembling the single comb of the other two species. The worker and honey cells are of the same size, but the drone cells are a little larger. They usually lie on the periphery or at the bottom of a comb, while the honey cells are near the top. In the modern hive, the frames which contain honey are above and somewhat separated from the brood combs. In a very thriving colony, the queen might wish to lay eggs in some of the honey combs. But by making the only access to these upper combs a hole which admits workers but is too small for the queen, the honey and the brood can be completely separated. A small amount of honey stored with the brood will be used as food by the bees.

It was early found that hives could be made in which some of the walls were of glass. These windows are normally kept covered with felt which can be removed at intervals when the bees' behaviour is to be watched.

The honey bee occurs in three castes, much more different from one another than those of wasps or humble bees. The males or drones are considerably larger and stouter than the workers and have greatly enlarged eyes which cover most of the surface of the head. The tongue is not well enough developed to be useful for gathering nectar, though the drone can feed himself on honey from the cells. He never does any work in the hive. He lacks the wax-producing glands and the pollen-collecting apparatus. The queen, also, is in some respects a "degenerate" or at least specialised type, for she also is unable to produce wax or to gather pollen or nectar. She is a little larger, especially longer, than the worker and has immensely developed ovaries which allow her to lay one to two thousand eggs a day and hundreds of thousands of eggs in her life-time. The worker has wax-glands on the ventral side of the abdomen, has a pollen-basket on her hind legs, and has mandibles better developed for work in the hive than those of the queen. Normally the worker's ovaries are small and non-functional and the little sac in which sperm are stored in the queen is rudimentary.

Occasionally, particularly in certain varieties, a worker's ovaries develop and she lays eggs which normally produce only drones. Very exceptionally, by a process which is not understood, some unfertilised worker eggs may produce new workers or even queens, especially in some African strains. As a rule, however, as in other bees and wasps, only the fertilised eggs of the queen can produce queens or workers. The honey bee is thus committed to a completely social life, since the queen cannot found a colony alone like a queen humble bee, while the workers cannot normally reproduce themselves.

DRONES AND THE NUPTIAL FLIGHT

The drones are produced right through the summer, the season of the honey-flow, and leave the hive only on warm days. Their single function is to fertilise the queens, of their own or of some other colony. Fertilisation takes place only in the air, a swarm of drones following the virgin queen as she flies upwards. The male is always killed in the act of fertilisation, since he can eject the sperm only by generating great pressure in his abdomen with the aid of muscles and of the fluid pressure of his blood. Under this pressure, the hinder part of his genital system is forced out and is left behind in the body of the queen. The pair then fall to the ground and the queen pulls herself away and returns to the hive, leaving the dead male behind. The remains of the male genital system, from which all the sperm will have passed into the queen, are removed from the end of her abdomen, some hours later, either by the queen herself or by workers.

This natural pairing is necessarily uncontrolled, and the male will belong to any variety which happens to be flying at the time. To improve their stock beekeepers have recently developed a process of artificial insemination by which carefully chosen queens and drones can be crossed. Sperm are removed from the male with a fine syringe. Then while the queen is held in a clip under a binocular miscroscope, her genital orifice is held open with small hooks and the sperm are injected into her. This makes selected high grade queens available to all beekeepers. To use such a queen, the old queen must be removed from the hive and the new queen introduced. Such a stranger would be attacked if precautions are not taken. Usually she is introduced in a small cage which protects her until she is thought to have acquired the nest smell. She is then allowed out into the hive.

At the end of the summer, when no more virgin queens are likely to be produced for some months, the drones are driven out of the nest by the workers and die of cold or hunger. They are probably never or very rarely actually killed or stung, but exclusion from the warm hive is just as effective.

While the characters of the drone are determined by his development from an unfertilised egg, the differences between the queen and the worker are known to depend on the diet which the grub receives. The queens develop in special, large, irregular queen cells which usually hang from the bottom of some of the brood-combs. These cells are built when required and are taken to pieces when they have been vacated; they are not used again like the other cells. All grubs receive the same food for the first two days—a substance known as "royal jelly" which is secreted by a gland opening into the mouths of the workers. It is not produced by queens or drones. Queen grubs receive this food throughout their life, and since the queen cells are frequently visited by the workers the supply of this food is always adequate. After the third day, the food given to the grubs of developing workers and drones contains an increasing proportion of honey. If a grub in a worker cell is transferred to a queen cell before the third day it develops into a queen, and *vice versa*.

About sixty years ago von Planta claimed that there were relatively simple chemical differences between royal jelly and the food given to worker grubs after the third day. More recent analyses by more refined chemical methods show that the nature of the food is much more variable than von Planta supposed, and the difference may be more in the quantity of the food than in its chemical composition; or at least that the chemical differences must be subtle and not yet properly understood. Nevertheless, the transference experiments seem to prove that diet is the essential factor in producing queens or workers.

The difference in diet also affects the time which is necessary for the adult to develop, for, while all eggs hatch on the third day, it takes the queen on the average thirteen, the worker eighteen and the drone twenty-one more days to complete development. These times seem to be related both to the richness and quantity of the diet received, and to the relative sizes of the three castes.

Queens are in general produced by the colony under two conditions: when it is preparing either for swarming, or for supersedure. Swarming is the normal method by which the colony reproduces, and in preparation for it several and in some varieties very large numbers of queen cells are constructed. Supersedure is the process by which a queen who is old or whose oviposition is falling off is replaced by a young, vigorous queen. In preparation for supersedure only one queen cell may be constructed and rarely more than three. A special case of supersedure occurs if the queen unexpectedly dies, and the rearing of a new one becomes necessary to meet the emergency. In this situation the queen will be reared in a " post-constructed " queen cell, that is one built from a worker cell already containing an egg or, more often, a grub less than three days old. After supersedure, the new and old queens may live and oviposit together for a long period without fighting. The intense rivalry between queens occurs only after swarming, or if two young queens emerge at about the same time during supersedure.

The honey bee differs from all other social species whose colonies reproduce by swarming, because it is as a rule the old queen which accompanies the swarm. Exceptions chiefly occur in secondary swarms, when the colony swarms more than once during the season. The prime swarm accompanying the old queen is made up of a high proportion of the older workers. Left behind in the hive is a correspondingly higher proportion of young workers, and several queens still in their cells but approaching the time for emergence. Normally, the first one to hatch becomes the new queen, and she destroys any others which emerge or which are in the adult stage but have not left their cells. The remaining queens cells which may contain grubs or pupae are then destroyed too, usually by the workers. The new virgin queen will have to fly out for her mating flight, usually when she is about a week old, and she begins to lay eggs two or three days later.

While some swarming is desirable in order to multiply the number of colonies, the bee-keeper must regard it as a wasteful process, particularly if it occurs more than once in the season. Swarming must be preceded by the rearing of a large brood to give a large population, and this uses up much food and energy and means that less honey will be stored. Moreover the workers in the swarm carry away large quantities of honey in their crops ; this enables them to secrete wax for the construction of new combs.

BEE ECONOMICS

The hive bee collects and stores five substances, nectar, pollen, propolis, balm and water, as well as itself producing wax. Nectar is obtained from a wide range of flowers by the foragers. An individual bee tends to keep to one kind of flower for long periods, probably as long as the secretion of nectar makes it an attractive source. She also tends to keep to one relatively small patch of the flower but she is capable of comparing sources, and changing to a new one if the nectar is more copious or more concentrated. Ribbands has recently analysed much of this behaviour by marking individual bees and watching what they do when collecting pollen or nectar. Some flowers, such as buckwheat, secrete nectar mainly at a particular time of the day ; the bees which visit it may restrict their foraging to this period, apparently spending the rest of their time inside the hive. By exposing dishes of syrup at particular times of the day it has been shown that the foragers have quite an accurate time-sense and can be trained to come for it at a particular hour. When the nectar is brought in it is handed over to a home bee which spends some minutes in concentrating it by sucking it in and out over the surface of the tongue before storing it in one of the cells.

Pollen is just as important to the bees as nectar, since it is their source of protein. Without it, the life of the workers is very much shortened and the brood cannot be reared. A considerable store of pollen is always kept through the winter, and this makes it possible for brood-rearing to begin early in the spring when fresh pollen is relatively scarce. A small quantity of honey is added to the stored pollen ; this mixture is " bee-bread ", and it has good keeping qualities.

Propolis is made from resin collected from various plants, especially trees. It is used to stop up any small crevice in the hive and to cover up objects which are too large to be removed. Some varieties of bees collect much more of it than others do. Balm is a particular variety of propolis, more liquid and transparent than usual. There is some doubt whether it is collected from particular sources of resin or whether it is secreted by the bees themselves. It is used to varnish the inside of the cells before eggs are laid in them.

Water is collected in quite large quantities, especially in hotter, drier climates. It is apparently used both to cool the hive by evaporation and to prevent the interior of the hive becoming too dry to suit the grubs. It is sometimes also used to dilute honey which has become too concentrated. The hive is maintained at a very steady temperature, only a little below that of our own bodies. The chief source of heat is provided by the activities of the bees and their brood, since the digestion of sugar automatically releases considerable energy. Workers regularly ventilate the hive by fanning their wings at the entrance. This cools it to a considerable extent, even if water has not been brought in and distributed, since the nectar loses much water as it is concentrated into honey. Ventilation serves to remove excess water quite as much as to reduce the temperature. It is the brood which is most sensitive to cooling and it always occupies a roughly spherical space in the centre of the hive, spreading through a number of the combs.

It is clear that all these manifold activities demand a considerable versatility in behaviour, and it is known, largely from the experiments of Rösch with marked bees in a glass-walled observation hive, that each worker tends to undertake a number of successive activities according to its age. During the active summer months the workers live five or six weeks and the succession of duties is as follows:

AGE IN DAYS	DUTY
0–3	Cleaning brood-cells and keeping brood warm
3–6	Feeding older grubs
6–14	Feeding younger grubs and the queen
14–18	Secreting wax, comb-building, cleansing hive
18–20	Guarding entrance of hive
20–40	Foraging

This schedule is not rigid. There is some overlap between successive duties, and probably not all bees reach the same physiological age in the same number of days. Still more important, the schedule can be widely modified to suit the needs of the colony. After swarming, for instance, there may be an unusually low proportion of the older workers and the foraging may then be done by much younger bees. Such a result can easily be demonstrated experimentally by making up a colony in which all the workers are of about the same age.

The key function is that performed between the sixth and fourteenth days, for this is the time when the workers are producing the royal jelly. This is a secretion of the pharyngeal glands which is passed into the alimentary canal just inside the mouth. This liquid, on ejection from the mouth into the cells, thickens into a pale jelly-like paste. Glands of the same type are found in all bees and wasps, but their function in other species has not been demonstrated. In the honey bee, the glands become active on about the sixth day and shrink and cease to secrete on the fourteenth, so that this age-group of workers is essential for brood-rearing. Apparently, the development of the gland can be a little hurried up in emergencies, and it is said that it may become functional a second time if older workers receive a diet rich in proteins.

The existence of a schedule of duties which is capable of modification for changing circumstances is characteristic of the social organisation of bees, but we have no idea, as yet, how this result is attained.

THE LANGUAGE AND DANCES OF BEES

The most remarkable advance in our knowledge of honey bees is due to von Frisch's discovery of what he has termed " the language " of bees. The first part of this work was published in 1923, but von Frisch largely dropped the investigation for twenty years. It was taken up again in Austria during the Second World War and led to discoveries so

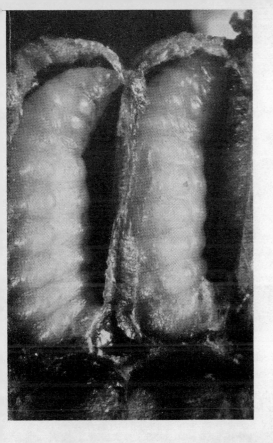

38. Full grown larvae of drone honey bees about to spin cocoons and to undergo metamorphosis into adults.

39. Full grown larvae of worker honey bees in capped cells spinning cocoons.

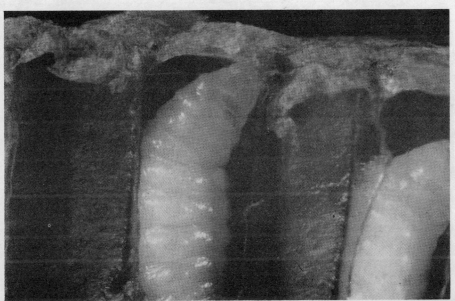

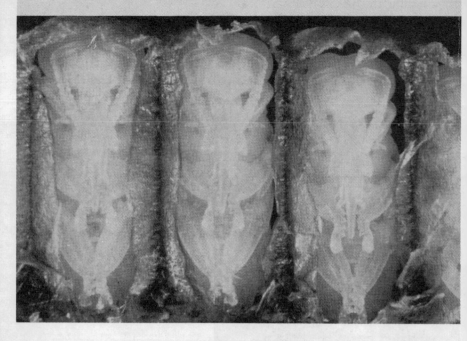

40. Worker pupae in their closed cells.

41. Worker bees on a comb whose upper cells are filled with honey. It is in groups such as that on the left that the dances take place (see Fig. 8, p. 114).

42 and 43. Research on bee behaviour. *Left*, marked bees feeding at an experimental station. *Right*, bees marked in various ways for individual observation. These photographs illustrate methods used in research on the senses and "language" of bees.

44. The honey bee. A comb with thimble-shaped "royal" cells in which fertile females develop.

45. Young worker honey bee emerging from the cell in which it was reared.

46. A honey bee swarm divided in two.

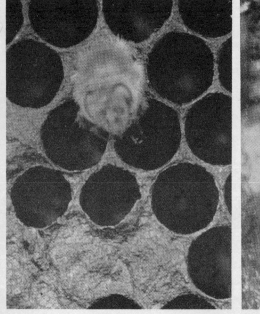

amazing that some biologists have found it difficult to credit them. At least the main outlines of the work have, however, been confirmed in the last few years by other observers, for instance at Rothamsted in England.

It has always been difficult to understand how such a highly organised system as a colony of any social insect could run without a general staff and with no means of communication. While no system of government has yet been discovered, very peculiar methods of conveying information have been revealed.

The principal experimental method has been to put dishes of sugar and water at points various distances from the hive. When foraging workers begin feeding at the dishes and taking the sweet liquid back to the hive, the bees are marked in various ways on their backs with coloured paints, so that each one can be recognised again. By watching the behaviour of the marked bees, both at the dishes and after their return to the hive, observers have discovered their methods of conveying information.

It is known that worker bees have glands at the end of the abdomen, between two of the overlapping plates of the upper surface. When the bees are feeding on a rich source of food these glands can be seen extruded, showing as small white patches. Von Frisch found that if he covered the extruded gland with a small patch of shellac, the dish did not in the next hour or so attract nearly so many new bees as did a similar dish at which the marked bees had not been treated with shellac. As the structure of the gland resembles that of a scent-producing organ, it is inferred that the dish becomes impregnated with a scent which guides newcomers to it.

Von Frisch further found that foraging bees, on their return to the hive, perform a sort of dance, either on the alighting board or on the comb. He found that if the source from which the worker is returning is no more than about 100 yards away, the bee does a "round dance", running round a circular track, first one way and then the other. Some of these bees then leave the hive, usually before the returned forager sets out again, and search for the source. Some of the workers in the hive closely follow the dance and may partly join in it. At these short distances the search is mainly a matter of quartering the ground in all directions. It was found by experiment that if a worker returned from a dish of syrup placed, say, at fifty yards north, then after the dance the issuing foragers were no more likely to find this dish than another one in a different place the same distance away.

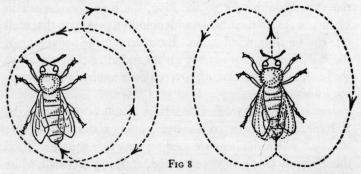

FIG 8

Dances of the honey bee. Left, round dance ; right, tail-wagging dance,

If, however, the original or "training" dish contained syrup scented with lavender water, then, if afterwards four dishes were put out, one scented and three not, the foragers all collected at the scented dish, even if it was exposed at a different site from the training dish. Thus, the round dance conveys the information that a rich source of food is to be found somewhere near the hive. If the food has a scent, as will be the case with most natural sources of nectar, this provides an additional clue which helps in its discovery, Finally, the bees themselves produce a scent which may give additional help to the foragers.

If the source of food discovered by a forager is more than about one hundred yards away, it performs a different dance on its return. This is the " tail-wagging " dance, and consists of running along a track shaped like a figure of eight with the part where the two loops join straight. As the bee runs along the straight piece she wags her abdomen. It was found that the further away the source of food was, the more slowly the dance was performed, but the more wags of the tail were made on the straight part of the run. The figures range from about eleven figures of eight in 15 seconds for a source at 130 yards, to four completed in the same time at 2,000 yards. The number of tail-wags in the straight run varied from two to three when the source was at 130 yards to ten to eleven when it was at 830 yards. These are average figures, and they are subject to some variation. Moreover, round about 100 yards transitions are sometimes seen between the two types of dance.

A relatively simple experiment demonstrates that the dances do convey information about distances. It is necessary to have a special observation hive with glass walls, so that the dances can be seen. Then suppose a source of syrup is exposed at 200 yards. After a time, scout bees find it and they, and later other bees, begin taking syrup back to the hive. The first dozen or so bees which come are caught and if necessary killed. Next a new source is exposed much nearer the hive and the number of new bees which come to the two sources in the next few minutes is compared. It is found that many more bees go to the distant source than to the near one, though in general with random searching near sources are discovered much more quickly. This is taken to show that the dances which the marked bees have been seen to perform have told other bees in the hive of the distance (and as will be shown in a moment) also the direction of the distant source. Thus until some bees have had time to convey information to the hive about the nearer source, the more distant one is much more visited. The number of visits to a source depends at first on distance but afterwards on whether the hive has been told about it. When told where to go, workers find a source much more quickly than they do by random searching.

The most extraordinary of von Frisch's discoveries was of the method by which bees indicate the direction of a source of food. The workers returning from a source may perform the dance on the horizontal alighting board just outside the hive. This is fully exposed to daylight and under those circumstances the straight part of the track of the tail-wagging dance points towards the source of food. With the round dance, the direction is not indicated, as far as is known. However, both dances are usually done in the dark, inside the hive, on a comb which is vertical. In this case, the vertical straight run of the tail-wagging dance indicates that the source is on the imaginary straight line connecting the hive with the sun. If on the straight run the bee moves upwards. the source is towards the sun, if downwards away from it, If the straight run is at an angle to the right or the left of the vertical, the source is at the same angle to the direction of the sun. This behaviour was observed in numbers of marked bees collecting food from known artificial sources placed in different positions relative to the hive and the sun. The direction of the dance varies somewhat on each occasion the straight run is made, and the average direction varies to some extent in different workers, and at different times of the day. In one experiment in which the direction of the dance was recorded for twenty-four dances by fourteen workers over a twenty-minute period, the angles recorded were as follows : 53° twice, 55° once, 60° seventeen times, 65° once, 67° once, 75° twice.

There is finally the problem of what the bees do when the sun is obscured by cloud. When the sky is completely overcast, no dance is performed, and in the dark the tail-wagging dance on a horizontal surface becomes completely disorientated. If, however, a patch of blue sky is anywhere visible, the tail-wagging dance is performed as usual, the direction being indicated with reference to the true position of the hidden sun. It has been suggested by von Frisch that this could be explained if bees were sensitive to the plane of polarisation of light, for it is known that the light reflected by blue sky is partially polarised, and that the extent of polarisation depends on the distance of the patch from the true position of the sun.

It should be explained that the light from the sun is emitted in rays made up of waves vibrating in all planes, that is to say unpolarised. But such light, when reflected from blue sky, is more or less polarised, that is the light is orientated so that the waves are all in one plane. Our own eyes are not sensitive to the plane of polarisation of light, though motorists know that a screen of " polaroid " which transmits polarised light will cut down the glare of the light reflected from puddles on the road or from snow. Von Frisch has very recently found evidence that bees' eyes can respond to the plane of polarisation of light, and that the direction of the tail-wagging dance can be altered if a polaroid screen is placed between the bees and the patch of blue sky which they are using to detect the position of the sun. In fact, when the polaroid is rotated through an angle the direction of the dance is shifted through the same angle.

The subject is, however, still very little explored and there are many details to clear up. Besides the honey bee, some ants also react to the plane of polarised light.

The development of communication in the honey bee is clearly an enormous advance in social organisation. Although it was claimed earlier that social life starts when any female looks after the young of one of her sisters, there is no comparison between the slight extension of normal nursing which this involves and the process of conveying topographical information. Moreover, there is no reason to think that we have more than crossed the threshold of an immense new field, since most of the more striking discoveries in it have been made since 1940. Butler has now shown that the workers are inhibited from constructing queen cells by a substance which they lick from the body of the queen. The substance is produced by the queen's mandibular glands and has been identified chemically as 9-oxodec-2-enoic acid by Callow and Johnston. Gary and Morse have been able to show that a queen whose mandibular glands have been surgically removed no longer inhibits the workers.

6

THE ANTS OR PISMIRES

In the ants we meet a group every member of which is social. All of them, except a few degenerate types mentioned in the next chapter, are found in three castes, often of very different appearance. These are the queen, the male and the worker; in some ants the worker may be of two types and when these are sharply different and do not intergrade, the larger one with a larger head is known as the soldier. As a rule the male is winged and often like one of the lower solitary wasps in appearance. He does not long survive pairing and can be found only during a short part of each season. The queen is usually considerably larger than the other castes and is also winged. The wings are, however, removed after the marriage flight. The winged females are thus found only for a short season, though the queen herself may be very long-lived, sometimes surviving for fifteen years. The workers are usually different from the queen both in colour and structure. They do not possess wings, and the absence of wing-muscles has led to a reduction in the size of the thorax and a simplification of its structure. Often the eyes are very small compared with those of the queen and in many species they are minute or absent.

Ants are much more flexible in their building habits than bees or wasps, and they keep their brood piled in chambers, not separated in cells. Most nests are built of soil or of small vegetable fragments, and they are more easily adapted to different habitats or to an increase in the size of the colony. Some ants add saliva to vegetable matter, to construct a stiff " carton ". A few use silk, which their grubs spin, to join leaves together. Usually, however, no scarce materials are required to make the nest. Perhaps one reason for the success of the ants has been that their nests do not depend on hardly won materials. They are at any rate more ubiquitous than any other group of social insects, and the only habitats on land in which they are not found are those of the Arctic. They are also much the easiest of the social insects to keep under observation in captivity. Many species can be maintained for years in small chambers of plaster of Paris with a glass lid. It is necessary only to keep them damp and to feed them on sugar and on occasional dead insects. Much of our knowledge of ants, especially of the early stages of the colony, depends on observations made on captive colonies.

There are good grounds for thinking that ants have led a social life for a much longer period than either bees or wasps. Certainly the castes were fully developed at the time the Baltic amber was fossilised, thirty to forty million years ago, and even long before this. We know this, because we find ants perfectly preserved in the amber. It is probable that their social organisation will be found to be more elaborate than that of the bees, but at the moment it is much less understood. Although it is not difficult to keep ant colonies in observation nests, no ant species has been as intensively studied as the honey bee. An immense amount has been recorded about the general natural history of ants, but most of this information is purely descriptive. That is, though we know in a general way what many species *do* under natural conditions, we do not know how they do it. A real understanding of ant behaviour would require much more investigation by means of experiment.

THE ANTS OR PISMIRES

ANTS AND OTHER ANIMALS

Let us consider the relation of ants to the aphides and mealy-bugs which provide much of the sugary part of the diet of many species. This is what Zimmermann says about the pineapple mealy-bug, a scale insect which causes very serious losses in the pineapple plantations of Hawaii.

> Mealy-bugs become established on the pineapple planting material during growth. These individuals persist on the future plant sets while they are partially dried after trimming and can, on occasion, even reproduce on this detached material. Hence, a newly planted field may have a large initial mealy-bug population. However, peculiar as it appears, most of these populations disappear either because of the lack of adequate attending ant populations or because of predator pressure or other causes. The serious infestation of a new field comes from outside, adjacent areas, especially old pineapple fields, waste or uncultivated lands. High infestations may build up from the outside within a period of six months. This movement is dependent primarily upon the action of the attending ant *Pheidole magacephala*. If the ant populations are low, the subsequent mealy-bug invasion of the fields will be slow and of low grade. If the ant could be eliminated, the fields would be largely free of serious mealy-bug infestation. Ants are essential for the proper develop ment of mealy-bug colonies, for they tend them, shelter them, protect them from parasites and predators and keep them clean from detrius which, when massed with honeydew, has a gumming-up and deleterious effect on the colonies. In short, pineapple mealy-bugs have a difficult time maintaining themselves unless ant-attended.

There is another side to this picture. According to Clausen, the date-growers of the Yemen have for centuries carried certain ants' nests to their orchards to protect them against other ants. In China, the citrus-growers collect nests of the red ant, Oecophylla, and put them in their trees, sometimes with bamboo runways from trunk to trunk. The ant is stated to kill many caterpillars and to drive away various beetles and bugs.

The fact that ants feed on the sweet, liquid excreta of aphides and mealy-bugs has been known for at least two hundred years. Linnaeus, indeed, said that aphides were ant-cows. Ants often keep certain species in their nests. Many years ago Sir John Lubbock showed that ants collect the eggs of the bean aphis in the autumn and in the spring plant them out again on the food-plant. El-Ziady and Kennedy have studied experimentally the effects of the Common Black Ant, Lasius niger, on the Black Bean aphid, keeping records of attended and unattended aphids. Some protection was given them from their enemies but the important effect was to make the aphids grow and reproduce more quickly. Fewer winged forms which would leave the plant were produced and the older wingless aphids wandered less. These effects seemed to be due to stimuli conveyed to the brain of the aphid and not merely to the removal of the honeydew.

THE FOOD OF ANTS

The methods by which ants provide themselves with food are analogous to three types of human culture recognised by anthropologists, namely, hunters, food-gatherers, and agriculturists. Many ants, especially those which in structure most resemble the non-social wasps, are purely hunters. They catch and kill insects or other small animals. As a rule, these are brought home whole and then cut up, to be divided amongst the members of the colony. Often all available prey of the right size are captured; less commonly, an ant may specialise on a particular sort of prey. The American Leptogenys, for instance, catches mainly wood-lice. Most of the hunters have a well-developed sting.

The food-gatherers collect either the seeds of various plants, or nectar from flowers, or the liquid excreta of aphides and scale insects. Some of them combine such collection with hunting, but others, such as the harvesting ants, live entirely on seeds, especially of grasses. In the harvesters the large-headed soldiers act as seed-crushers for the colony. The large head contains the very powerful muscles which work the mandibles.

The most highly developed agriculturists are the leaf-cutting ants. They subsist entirely on a special kind of fungus grown on the fragments of leaves which they carry into underground chambers. The queen of these ants takes a little pellet of the fungus with her on her marriage-flight, so that the food is passed on to each new nest.

Some of the ants which feed on the excreta of aphides can also be classed as agriculturists. The common yellow ant, Lasius flavus, which makes mounds in fields, obtains almost all its food from aphides living on the roots of plants in underground chambers in or near the nest. These might well be compared to herds of cattle, except that we do not yet know how far the ants manage them in the way in which man does his domesticated animals.

Some of both the food-gatherers and the agriculturists have the sting greatly reduced, and in their attacks on their enemies squirt the poison instead of injecting it with a stylet. It is not clear what they have gained by the change, though some of the most common and successful species retain the sting only in a rudimentary form.

We may now consider in more detail the three main ways by which ants feed themselves. A great many ants subsist partly by hunting, as even civilised man does to a limited extent—by fishing, for instance. The wood ant, which often makes large mounds in pine-woods, brings in large quantities of small insects such as caterpillars. It may also find small dead insects which are worth collecting. Ants are always a serious nuisance to solitary wasps whose prey is stolen while awaiting storage in the wasp's nest. But the wood ant also collects large quantities of aphis excreta. Okland has calculated that a large nest in Sweden may collect as much as 20 lb. of dry sugar in a season; this means a very much larger quantity of the liquid excreta. As a rule, hunting workers act singly, so that the prey are no bigger than one ant can overpower and carry by itself. In some ants, however, the workers co-operate to overpower much larger insects, such as big beetles or even small birds.

Hingston has described how this is done by an Indian ant, Oecophylla smaragdina. One of a number of hunting workers happens to find a beetle and seizes one leg. It is too powerful for the one worker. There is a struggle which soon attracts others. Each one seizes some projecting part and pulls, and soon the beetle is splayed out with all its members pulled in different directions. The ants continue a steady pull, if necessary for half an hour, until the prey is dead. The workers then co-operate to bring it home; most of them pull the legs on one side, but a few remain on the other side to steady it. In this way it will be dragged home even if the nest is high up in a tree.

There is one group of ants, the Ponerines, which are pre-eminently hunters. Very few of them collect aphis excreta or are attracted to sweet substances, though a few of them feed, at least partly, on seeds. In structure the Ponerines are the least specialised of the ants, in many ways most like what we imagine the ancestors of the whole group to have been. Both in their addiction to hunting living prey, in the small size of their colonies, and in other features they seem to be more primitive than other ants. They are now relatively rare and most live in retired situations, especially in the tropical rain forests. Like the marsupials, which are primitive mammals, they are specially numerous in Australia. Here one of the dominant forms is Myrmecia, the bull-dog ant, which is by no means retiring. The workers of some species are nearly an inch long and are capable of jumping a foot at a time. Froggatt describes how, if their home is approached, they come jumping out one after another like a pack of dogs. As they can bite and sting severely the nest must be treated with great respect. Another Ponerine, Lobopelta, is a very active hunter and tends to specialise on termites. Hingston has described how an Indian species sallies out in dense armies to attack termite nests. Some of the ants enter the nest and begin to kill termites, throwing them out or bringing them to the entrance. Other ants act as porters and carry the prey back to the ants' nest. All who have observed these and other raiding parties of ants are irresistibly reminded of a skilful military operation.

THE DRIVER AND LEGIONARY ANTS

This type of behaviour reaches its highest development in the driver ants of Africa and the legionary ants of South America. These together form a distinct section of the Ponerines and have much in common. The workers of the drivers are always blind, whereas the legionaries usually have rudimentary eyes. Some kinds are almost entirely subterranean, others merely nocturnal. One of the legionaries builds a tunnel of grains of sand as it goes along and in this way passes in shelter from log to log, looking for beetle-grubs, etc. Others especially in the tropical rainforest, are fully diurnal and easily observed, as I have myself proved in British Guiana. A raiding army covers an area of several hundred square feet, and as one approaches it one hears the characteristic chirping of the ant thrushes. These are birds which have the habit of catching the numerous insects trying to escape the ants; the birds themselves escape both by their size and mobility. One can, in fact, stand still in the middle of a raiding army and come to no harm if one moves on every minute or two. But there is no such escape for any insect or small animal which cannot fly away. Wasps' nests, for instance, are deserted and the whole brood is captured.

The African driver ants are even more formidable and will kill even pigs and fowls if these are enclosed. They have been known to devour a large python which was so gorged after a meal that it could not escape. These ants will enter houses and clear them entirely of the vermin, such as cockroaches and spiders. Fortunately, human food other than meat does not attract them, so that their visit is on the whole beneficial, at least in retrospect.

Most driver and legionary ants construct only temporary nests. Often they occupy some natural hole or crevice, such as may be found at the roots of a large tree. Some of them have the faculty of building a structure by using the bodies of some of the workers themselves. By hanging onto one another long chains can be formed, and the chains may be arranged to arch over and enclose spaces. In this way a temporary living nest is produced, inside which the queen and brood can be protected and in and out of which the other workers will be seen passing. The temporary nature of most of these colonies appears to be the inescapable result of their mode of life. Ants which occur in such large numbers and which live exclusively by hunting soon denude any one area of food and have to move on. There are many accounts not only of the ordinary predatory raids but of the shifting of the whole colony when all the brood is carried to a new site.

Some idea of the physiological background to the restless activity of these ants has been obtained recently in America, by Schnierla. There are four types of raiders—subterranean, arboreal (mainly), terrestrial swarm-raiders and terrestrial column-raiders; Schnierla's observations were made on the last two types. They differ in the way their raids are organised; the first makes a swarm which fans out over a considerable area, the second moves in much more compact, narrow columns.

The ants have a regular alternation of two phases of activity: one which lasts about three weeks during which the nest or "bivouac" is stationary and only a few raids are made, and one which lasts fourteen to seventeen days in which the bivouac is moved almost every day and during which there are several raids each day. This alternation of relative inactivity with great activity depends on the oviposition-cycle of the queen. She goes through a short phase in which her abdomen becomes greatly distended and she then lays about twenty thousand eggs in a few days. When these hatch and while the brood is growing, the colony is very active and raids are frequent. When the grubs are nearly full grown about two quarts of insect prey are brought in every day. As these grubs begin to spin their cocoons, raiding dies away rather suddenly and the bivouac becomes stationary. Activity begins again just before the new lot of ants emerge from the cocoons. The regularity of their behaviour depends on the regular oviposition-cycle of the queen and on the way in which almost all the brood keep in step with one another during their development.

It is still not understood how the queen maintains a regular cycle in egg-production. It is clear, however, that though a stationary colony consisting of eighty thousand or more ants could not feed itself for long by raiding, it is not the shortage of food which directly controls their movements. It is rather that the need of food for the growing brood stimulates raiding and that when raiding is active the bivouac is frequently shifted. On each occasion all the brood have to be carried to the new site.

GATHERERS OF MIXED FOODS

Just as most ants do at least a little hunting, so most of them will collect other foods of various types, especially the sweet excreta of aphides, scale-insects, mealy-bugs and leaf-hoppers. Some ants, like the common red stinging ants of the genus Myrmica, are hunters as well as collecting some nectar from flowers and some excreta from various insects. At the other extreme, the common yellow mound-building Lasius flavus subsists almost entirely on the excreta of root-aphides, but also takes some dead caterpillars, etc. I once saw on top of an ant mound a large dead caterpillar completely surrounded by fifty or sixty ants dismembering it. From the great excitement which this ant exhibits if given animal food in an artificial nest it seems certain that this is an important supplement to its sugary diet. The relative importance of different articles of diet to the various kinds of ant is very little known, but it is clear that the balance is different from genus to genus.

None of the British ants builds elaborate shelters above ground to shelter aphides or mealy-bugs, though they may build up earth round the base of a plant which these insects inhabit. Many exotic ants, however, build elaborate structures around the suppliers of their sweet food and seem, at least to some extent, to protect them from their enemies No doubt the shelters are usually built round a natural colony of insects which the ants happen to have discovered, but there is a little evidence that they sometimes drive insects into their sheds and prevent them from escaping from them. Hingston describes an Indian ant, Polyrhachis, which builds

47. A worker ant (Formica fusca) with aphides. Gentle tapping of the abdomen often makes the aphid produce a small drop of sweet excreta.

48. Nest of wood ant (Formica rufa). The mound is three feet high and eight feet in diameter.

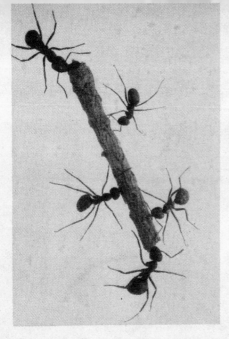

49. Worker wood ants (Formica rufa) combining to carry a pine twig to nest.

50. Nest of termite (Amitermes vitiosus) in Australian scrub.

51. Fertile female termite, or queen—nearly all swollen abdomen. Compare with Fig. 12 on page 174.

its shelters from varied pieces of debris which it spins together with the silk obtained from its own grubs. These shelters may be only an inch or two long or as long as a foot or more. They surround part of the stem or a few twigs from which the food-supplying insects suck the juice. This particular ant often keeps leaf-hoppers of the family known as Membracids, sheltering the adults and the early stages, both of which but especially the young ones supply sweet excreta. Hingston describes ants poking with their heads at a Membracid to make it walk along a branch until it entered the shelter. He also saw ants driving them back into the shelter when they escaped through a hole torn in it by the wind. They also seemed to prevent them from leaving by the entrances which the ants themselves use. A great many ants behave in a more or less similar way to Polyrhachis, but the exact meaning of their behaviour will be understood only when many experiments have been carried out.

HONEY-POT ANTS

One well-known and extraordinary development of this type of behaviour is shown in the honey-pot ants. They all occur in the drier parts of the world such as the southern U.S.A., Australia and South Africa. Here there is probably a need to store a reserve of food to tide over periods of scarcity, just as the honey-storing wasps, Brachygastra, do in Mexico. The best-studied species is the one which occurs in Colorado: this ant stores in the bodies of certain workers the sugary liquid which it obtains from aphides, scale-insects and from the surface of an oak-gall. These living storage-tanks are known as " repletes " and have their abdomens so greatly swollen that the hard plates which normally form a continuous armour are widely separated by the distended membrane connecting them. The repletes hang from the roofs of special underground chambers in the nests and are almost incapable of movement. If they happen to fall from their perch they are unable to get up again. They are fed by other workers who also bring the liquid which distends them. The repletes start life as ordinary workers, but before their skeleton has fully hardened they are given so much liquid food that their crop distends their abdomen and this condition becomes permanent, lasting for months or even years. Though the repletes form only a small proportion of the workers, there may be two hundred or three hundred in the nest. The repletes give up part of their store when solicited by other workers, and it is presumed that they can thus keep the whole colony going during a period of prolonged drought. It is a matter of common observation that many ants when returning from a visit to plant-lice have the abdomen considerably distended, but normally it contracts again as soon as they have shared out the food. This facultative distension has been much exaggerated in the honey-pot ants and has become permanent in the repletes.

HARVESTING ANTS

One group of food-gathering ants has been familiar from remote antiquity, namely the harvesting ants. It was to them that Solomon directed the attention of the sluggard. They are found in many of the drier parts of the world, especially round the Mediterranean basin. Many ants make some use of seeds as a source of food and the British red ant, Myrmica, already mentioned as a hunter and a collector of aphis excreta, has also been observed bringing cornflower seeds into its nest. Many seeds have a small part which contains more fat than the rest, and this is especially eaten by ants. Some botanists have, indeed, maintained that the fat-reserve of the seed has been evolved to make it attractive to ants, just as the nectar makes flowers attractive to bees. The ant thus acts as an agent in seed-dispersal. Whether or not this is true, it seems that ants do play some part in spreading plants such as gorse which have heavy, edible seeds. In this example, the ants eat only part of the seed; the seed consequently remains capable of germination.

The true harvesting ants live mainly on seeds, especially of grasses. Their behaviour has an extraordinary analogy with that of the honey-pot ants, for they store the seeds in underground chambers and use them as a source of food during periods of drought. Tracks leading to the nest often look like well-marked roads, since many species of these ants forage in large companies. Usually the seeds are brought in enclosed in the chaff, and this is removed inside the nest and brought out again and thrown on a rubbish heap which forms a sort of ring round the nest. In some species the nest may be a large mound, sometimes as much as 20–30 feet across and penetrating 6 feet or more into the soil. The entrance may be large and lie in a crater made up of coarse soil pellets.

It is well established, and was indeed known to the ancient Romans, that if the seeds get damp in their underground chamber, the ants bring them out and spread them in the sun to dry. Later they carry them in again. They may also bite off the part of the seed called the radicle to prevent them from germinating. In spite of these measures their stores do sometimes germinate; they are then thrown away on the rubbish heap. This gave rise to a story that the ants planted seeds to produce a crop near the nest for the following year. Although it is true that rejected seeds often germinate and grow in the rubbish heap, and that the crop of seeds thus produced may be harvested, there is no reason to suppose that this is anything more than an accident. Many of the harvesting ants have a special caste of large workers with enormous heads. These are supposed to act as seed-crushers, making the food available for the smaller ants who do all the collecting.

LEAF-CUTTING ANTS

The most elaborate agricultural operations are undertaken by the leaf-cutting ants. These grow special fungi on fragments of leaves or flowers which they carry into their nests and store in underground chambers. Leaf-cutting ants are found in America, from Texas to Patagonia, but are most numerous in the tropics of South America. There are many species, and there is a considerable range in structure and behaviour. It will suffice here to describe the largest and most specialised type, which belongs to the genus Atta.

In most parts of South America nothing is more familiar than the processions of the leaf-cutter or parasol ants. Each worker is carrying a fragment of leaf or flower, held upright over its head, " like Sunday-school children carrying banners " as the Rev. McCook remarked in one of the early accounts of their habits. In many districts they are a serious menace to agriculture, since they strip all the leaves of plantations, often seeming to prefer the non-native vegetation grown by the farmer. The colonies are of very large size, with well-beaten paths radiating in all directions from the nest. The nest consists of large irregular mounds of earth several feet high and 20 or 30 feet across. The entrances are as large as rat-holes, and there often seem to be long underground passage-ways leading to points more than 100 feet from the nest. The underground part of the nest extends downwards for 10 feet or more, and contains a large number of fungus-chambers connected by wide galleries. The chambers are quite large, perhaps a foot or more high and 1–4 feet wide. They may be connected to the surface by ventilating shafts which maintain suitable conditions of moisture, etc.

The leaf-fragments brought in are chewed up and placed in the chambers where a special fungus grows on them. This is not a mere mould growing on decaying vegetable matter, but a particular, very peculiar kind of fungus which seems to be different for each sort of leaf-cutter. Indeed the fungus is so peculiar that it is difficult to place in the botanical system. Much of the nutriment of the fungus is provided by the liquid faeces of the worker ants, which are added to the leaves; growth of the fungus penetrates all through the mass of leaves and develops a number of little knobs, almost like a cauliflower head, and on this all stages of the ants feed, exclusively. Some of the youngest grubs may be fed on food regurgitated by the workers, but the larger ones lie about amongst the fungus on which they feed themselves. The workers, as in so many ants, vary greatly in size. The smallest ones stay in the nest, weed the fungus gardens of moulds, and nurse the grubs; the middle-sized ones, which are the most numerous, cut and collect the leaves; the largest ones with very big heads and mandibles guard the entrance to the nest.

When the female flies off on her marriage-flight, a small pellet of fungus and debris is carried in the pocket which is found in all ants beneath the mouth. After mating, she sheds her wings, excavates a small underground chamber, and ejects this pellet onto the floor. The fungus soon begins to grow and is nourished with her liquid excrement. A piece of the fungus is taken in her jaws from the growing sheet and held against her anus from which a drop of clear yellowish fluid is ejected. It is then replaced in the young fungus garden and this may be repeated every hour or so. Thus this manuring of the fungus is not an accident due to living in a small enclosed chamber, but is an elaborate, instinctive performance.

At this time, also, the queen begins to lay eggs. Some of these are eaten and probably help to provide food for the fungus. Others hatch and others again may provide food for the youngest grubs. Soon after the first workers hatch, the colony begins to collect vegetable fragments or to cut leaves.

TYPES OF NEST

Ants' nests are in many respects simpler than those of bees or wasps, and this may give the colonies an advantageous flexibility. Even so, there is a considerable variety in the types of nests, some of which are remarkable structures. The commonest type is built in the ground, either under a log or stone or with a mound raised over it; more rarely there is almost no external sign. The mounds are built by the ants, bringing up the earth a few grains at a time, especially in the early summer. Such mounds in England are often confused with mole-hills, but they are made of much finer earth and the ants can be found by digging into them. By tunnelling in all directions, and leaving pillars to support the roof, they are able to make as convenient a nest as a bee or wasp, but in a much simpler way and with less work.

The larger mounds certainly take a long time to build, and many of them must be twenty years old or more. Since queen ants have been kept in captivity for more than fifteen years, and since replacement of the queen is possible as in the honey-bee, the colony can live for long periods without difficulty. According to whether it is dry or wet, hot or cold, the workers shift the brood about to different parts of the mound, to give them the right conditions. Some ants are said to have two nests, one for the summer and another, in a more protected situation, for the winter.

Other nests are less simply constructed—for instance those made of carton. This is a product of vegetable matter, usually wood-particles, cemented together with saliva. It ranges in texture from soft, leathery or friable material to something which is fairly tough and may be very like wasp carton. Some British ants build carton nests inside rotten trees, but many tropical species build large external nests which hang from the branches. There may be a number of chambers, hanging from different branches and connected

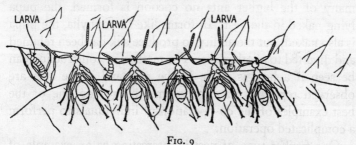

FIG. 9

Brigade of Oecophylla smaragdina workers drawing edges of leaves together while other workers bind them in place with the silk spun by the larvae (after Doflein).

by runways, so that most of a large tree is occupied. Other nests are more like a series of purses, each under a different leaf of the tree, so that the separate chambers are very numerous.

The most remarkable of the tree nests are those of the species of Oecophylla, already briefly mentioned. The way the nest is built has been watched by a number of observers and the same method has been seen in several other rather different kinds of ants, in other parts of the world. Several ants pull the edges of two leaves together; towards the tip of the leaves where the gap between them is wider, several ants form a chain, each one holding onto the one behind with its hind legs. The number of ants and the length of the chain is adjusted to the width of the gap at different points along the leaf. When the edges of the leaf are drawn together, another ant comes along bringing one of the larger grubs in its mouth. With the silk which is emitted from the lower lip of the grub, the leaves are sewn together by a regular criss-cross of threads. Eventually enough silk is produced to close the gap with a solid sheet.

This silk is the material which in Ponerine ants serves to form the cocoon when the grubs change into the pupa. In many of the higher ants no cocoon is formed, the pupa lying naked in the nest. In forms like Oecophylla, the pupa is also naked, but the ability to produce silk has been retained and diverted to another use. The method of construction can be seen if a hole is made in the nest and if the ants are observed repairing it. This seems to provide one of the best examples of the co-operation of individuals to perform a complicated operation.

One further type of nest is interesting as an example of the adaptability of ants. Many species build small nests in natural cavities in plants, such as hollow stems or empty oak-apple galls. A number of plants have natural preformed cavities which are regularly used by ants. Such are many species of Acacia, in which there are hollow stems or large hollow thorns ; other plants have swollen, hollow bases to their leaves. Certain kinds of ants are regularly and exclu-sively associated with such plants, nesting in no other situa-tion. It has been suggested that the plants have evolved in this way in order to attract the ants ; the ants in turn are supposed to protect them from the other insects which would feed on their leaves. There has however been a good deal of evidence that the leaves of such plants are no less eaten than those of plants not inhabited by ants, and the question needs much more study.

THE ANTS OR PISMIRES

HOW COLONIES ARE FOUNDED

The normal method of founding new ant-colonies is by a single queen after her marriage flight. These marriage flights often take place over wide areas simultaneously. One day during July 1951 the common black ant, Lasius niger, was swarming over much of the country between Ascot and Ealing, near London, a strip about 20 miles long. Such observations are not at all uncommon, and millions of queens and males are involved. It must lead to more crossing between different nests than would occur if each one swarmed on a different day. It is not known what conditions determine the time of swarming, though it is often during warm, thundery weather. Just before swarming the workers are seen crowded round the exit-holes of the nest, and appear to be restraining the young sexual forms from coming out too soon.

Usually the queen is very much larger than the male or worker. After mating, she falls to the ground, breaks off her wings at a line of weakness near their bases, and hunts for a crevice or digs a small cell under a stone. Here she lives without food for nearly a year, being sustained by food reserves derived from the breakdown of her wing-muscles. Only in the Australian bulldog ants and in some of the other Ponerines is she known to go out and forage for food during this period. When she starts egg-laying some of the eggs are eaten, so that however many she lays few survive. The small number of grubs which she raises are fed on her saliva, or on fragments of eggs or of other grubs. Eventually a brood of unusually small workers is produced, and from then onwards the queen can spend all her energies on egg-laying. This method of colony-foundation is possible only when the queen is much larger than the workers; otherwise she would not

---

contain enough food-reserves to rear the first brood. Probably the Ponerine ants, in which the queen is still relatively small and forages for food before the first workers emerge, retain a more primitive condition.

One may suppose that the cloistered cell of the majority of the queen ants has turned out in the long run to be less risky. It does, however, put a considerable strain on the colony, since it has to produce large numbers of big queens. Accordingly several modifications of the usual method are known. In some the size of the queen is reduced, and this entails a change in the method of founding colonies. Sometimes the new queen is received back into her own or a nearby colony. With many small queens the egg-laying rate can be even higher than when there is only one large one. Moreover the colony is potentially immortal: its survival does not depend on the life of a single queen.

The Brians have recently shown that the common British ant, Myrmica rubra, occurs in two forms. In one the queens are larger than the workers and colonies are founded by single females. The other has relatively small queens which go into the parent or some other nest after fertilisation. Ant colonies of this second sort may multiply by " budding ", either through the simple division of a very large and straggling colony or by a group of workers and queens seeking a new site. The co-existence of several laying queens does not seem to be an easy achievement for most species : jealousy of other egg-layers is much more characteristic. Moreover, budding is hazardous because the species cannot spread so rapidly or so widely.

The commonest way round this difficulty seems to be the type of behaviour known as "temporary social parasitism ". The small queen finds her way into the nest of *another* species, and succeeds in displacing the rightful queen. The alien workers bring up the early broods of the intruding queen, and as they die off and are not replaced the colony gradually becomes a society of the intruding species only. This type of behaviour seems to lead in the end to more and more pronounced parasitism of the sort which will be described in the next chapter. An example is the American ant, Formica consocians, which invades the nests of F. incerta; the British wood ant, F. rufa, has been said to have similar habits but perhaps more often buds off new colonies.

An entirely different method of solving this problem has been evolved amongst the driver, legionary and a few other kinds of ants. In them, the ant which lays all the eggs is wingless, blind and has a thorax formed like a worker rather than a queen. It is supposed that in these species the original queen has entirely dropped out, and that her functions have been taken over by a special type of egg-laying worker. Since this acting queen has no wings she mates on the ground and the species spreads only by the wanderings of the whole colony. This arrangement is not found in many kinds of ants, and probably has the same disadvantages as ordinary budding, especially a loss of mobility for the species.

THE CASTES OF ANTS

As a rule, in any one species the male and female are rather constant in form and colour, while the workers are much more variable. In the common, yellow, mound-making ant, the workers vary in size only to a moderate extent. In the wood ant the variation in size is much greater, and there is also a good range in the amount of black markings.

In many non-British ants the variation or polymorphism is much more marked. In driver ants or leaf-cutting ants the biggest workers are many times the size of the smallest ones. Moreover the shape of the mandibles, or of spines on the thorax, may vary in a marked fashion. Finally, in some groups, such as the large genus Pheidole, the workers are found in two types with no intermediates, the small workers with normal heads and the larger soldiers with disproportionately large heads. While some variation in size might be expected in any species, especially if there is any likelihood of competition for food amongst the brood, the variation within the leaf-cutters, for instance, seems far more than would be expected from such a cause. Besides, the grubs which give rise to the queens are brought up in very similar way, yet the queens vary little in size. Where the workers are dimorphic, i.e. of two discontinuous types, there must be some special explanation. However, it might be argued that the females of nearly all ants are dimorphic, developing either into queens or workers. The existence of the two sorts of worker in Pheidole might then be regarded as no more than a special case of the more general problem of female polymorphism. If we could explain why it is that some grubs produce queens and others workers, the variation within the worker caste might also be understood.

The question of what determines the development of a queen or a worker is one about which a debate has gone on for years; it is still not all settled. The answer is not necessarily the same for every kind of ant, though it is usually assumed that it is likely to be so. There are two main sorts of theory. It is suggested first, that the queen lays two or more sorts of eggs (genetic theories); second, that originally similar grubs are given different quantities or types of foods (trophic theories). In the midst of disagreement, there are at least two certain facts: that the differences between males and females (or workers) are genetic, and are due to the laying of two sorts of eggs; and that by starving ant-grubs at least some variation in the size of queens or workers, running parallel with the simplest kind of natural polymorphism, can be produced experimentally.

In the past, the chief support for the trophic theories has been derived first from what is supposed to occur in bees and wasps ; in these caste-production has been supposed to be entirely due to differential feeding. Secondly, in most ants the polymorphism of the workers is mainly a matter of size. Species with a sharply discontinuous soldier caste are few. Thus the most usual type of variation within the worker caste is very much what would be expected from an unequal division of food. It is known from observation nests that worker ants may give exceptional attention to grubs which happen to be larger than the others, so that a slight natural variation in size might be accentuated by the behaviour of the nurses.

Recently, some experimental evidence has been obtained which supports the trophic theory of the origin of castes. Goetsch and Gregg have studied the relation of food supply to the production of soldiers in Pheidole. This is an ant in which the large, swollen-headed soldiers are sharply differentiated from the small workers; no intermediates between them are found. Goetsch showed that a group of grubs given honey or some other liquid food turned into workers, whereas a group provided with pieces of dead insects produced one or more soldiers ; the number depended on the quantity of solid food. Liquid food is taken into the worker crop and then some is regurgitated for each grub individually, so that each one gets a more or less similar share. Solid food, on the other hand, is merely put near the grubs, so that one or more of them, by reaching it and feeding themselves, get a much bigger share than the others. This solid food must be received during the first five days of life, otherwise the grub irrevocably becomes a worker. The feeding on the fourth and fifth days seems to be most important, and there seems to be a short period when the grub can be most readily diverted from a worker into a soldier. The first group of eggs laid by a young queen is, however, incapable of producing soldiers even if the eggs are removed and given to an older colony for rearing. Soldiers can be produced only from eggs laid in the second or later batches of the queen.

In experiments by Wesson, on an American species of Leptothorax, an effect of food on the production of queens was shown. Grubs from one colony were made up into two groups of forty-four and put with workers in artificial nests. One lot were given abundant food, the other just enough food to allow growth and to prevent cannibalism. The results were as follows :

	WELL FED	POORLY FED
Initial number of grubs	44	44
Died during the experiment	6	9
Formed pupae	38	35
Produced males	3	2
„ queens	32	10
„ workers	3	23

It seemed, however, that food supply was not the whole explanation, since grubs present during the summer did not respond to such treatment. Only grubs which had over-wintered were responsive in such experiments.

Brian has very recently found some confirmatory evidence in the British ant, Myrmica rubra. He finds that the grubs of this ant all pass the winter in the third moult stage, though they vary greatly in size ; the biggest ones are ten times as heavy as the smallest. In the spring, when the colony wakes up to life, the biggest ones develop into queens, the rest into workers. If the ratio between nurses and grubs is altered experimentally the proportion of queens and workers produced can be similarly altered. With one nurse for each grub, provided enough food is given to the colony, all may become queens, whereas with one nurse for ten grubs, all will become workers. A similar result can be obtained by partial starvation of the half-grown grubs and in this way, in certain conditions, intermediates between queens and workers were obtained.

These experiments have not yet been described in full, and it is not known why the grubs vary so much in size at

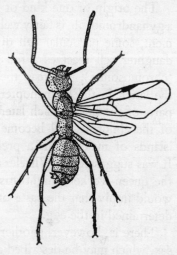

FIG. 10

Male-worker mosaic of the ant, Formica sanguinea, found at Bewdley, Worc., 20.vii.1909 (Bondroit). Right half male.

the beginning of the winter. There is always the fundamental difficulty that worker ants may recognise the caste of newly hatched grubs although we cannot do so. This might lead them to treat each type of grub in the appropriate way. On the other hand, in the leaf-cutter ants, as Wheeler pointed out, fungus appears to be almost the only food of all grubs which have access to an almost unlimited supply, and yet the workers are highly polymorphic.

What has been regarded as the best evidence for the

genetic origin of castes comes from the study of various abnormal individuals. Very occasionally specimens are found which may be called sex-mosaics. These have part of the body female, part male, with the areas often sharply marked off from one another. Where the male and female are different in size, shape and colour, such mosaics are asymmetrical and look very peculiar. Analogous mosaics which are male-worker, male-soldier, queen-worker, soldier-worker and soldier-queen, have also been found.

The origin of one kind of male-female mosaic, known as a gynandromorph, is fairly well understood. During development, some particular cell divides abnormally, so that one daughter cell receives a full set of the hereditary material (chromosomes), whereas the other gets only a half set: the latter, in the Hymenoptera, results in a male. All those parts of the body which later develop from the subdivision of the abnormal cell become male, and form an island or islands of maleness in a predominantly female individual. It was suggested by Wheeler that the other mosaics, such as the queen-worker, might arise in an analogous way. This would imply that the caste of each individual is genetically determined in the egg.

There is, however, another type of sex mosaic, the inter-sex, which may be described as a mosaic in time rather than in space. In the gypsy moth, for instance, every cell of the body produces substances which influence development in the direction of either maleness or femaleness. If the physiology of development is disturbed, there may at first be an excess of the female-promoting substance but later of the male-promoting one. The parts of the body which in development reach their final form first will then be female, while those which develop later will be male. The mosaics so produced tend to be of a type in which male characters chiefly appear in the head region and female characters in the more posterior parts. Whiting has suggested that in ants the mosaics are really intersexes or intercastes, and that the second type of anomaly at least might be due to a change in nursing behaviour during development. This is in fact supported by Brian's observations. It cannot, therefore, be claimed that the existence of caste-mosaics proves that the castes have a genetic origin. To carry the argument further it would be necessary to study the order in which the parts of the ants' external skeleton are laid down, and to test whether after such treatment as starvation the later developed parts show any change in their caste character.

No clear picture can as yet be given of the type of hereditary mechanism which could be operative in the more polymorphic type of ant. Some scheme, like the rather theoretical one already mentioned for stingless bees, might be imagined. There are two special features about the social life of ants which might make some such theory more tenable. The possibility of the differential feeding of grubs whose caste the worker ants had recognised at the very beginning of development, has already been mentioned. It is further possible that some of the egg eating which is common in ants, especially in young colonies, may also be differential. Thus, only worker or queen eggs might be allowed to survive at certain seasons, though both types of egg might be produced at all times. This would explain why the supposed hereditary mechanism, which would be expected to produce all castes in fixed proportions, does not in practice do so, the castes being produced rather according to the needs of the colony.

There is another process which operates, not only in the ant colony but in any social animal. The unit whose efficiency determines whether the species shall survive or become extinct is the colony rather than the individual. An individual which is useful to the colony may survive though it would be quickly eliminated in a solitary species. This happens to a considerable extent in man. In civilised societies, many members are supported who contribute only very indirectly to the provision of the food and shelter necessary for life. Others who contribute nothing are enabled to survive because our social behaviour benefits the whole species and not merely the bread-winners. In an ant-colony there is an analogous situation: the worker which is sterile, or capable only of producing male offspring, is a good example of a type which could not survive apart from the colony. Some of the more fantastic types of soldier ants seem to be an even more extreme example of the same thing. They might be described as freaks for whom, during the evolutionary process, a use has been found, just as the circus has found a use for dwarfs.

COMMUNICATION

The social life of ants involves extensive co-operation between individuals for the good of the whole colony. The spinning together of leaves by Oecophylla is a striking example, but a great deal of ant behaviour is of the same type. To say that each individual is reacting instinctively to the needs of the colony is a description rather than an explanation of what is observed. Some ant behaviour might be better understood if ants were known to have means of communication of the type described in the honey bee. Although, so far, nothing of this sort has been discovered, this does not prove very much. Facts as strange as those revealed by von Frisch for bees may yet come to light. Meanwhile, the little that is known about communication and the related topic of path finding may be briefly summarised.

Some ants living on trees tap the leaves with their heads as a danger signal. When a number of workers do this simultaneously a clearly audible sound is produced ; but this can hardly be more than a generalised signal, like the blowing of a policeman's whistle. Other ants are able to produce a slight sound by rubbing together the roughened surfaces of two overlapping parts of the abdomen. This appears to be used as a danger signal, but it seems unlikely that it conveys any very precise information : some information is also conveyed in the exchanges of food which are such a regular feature of the ant-colony. A single worker bringing in a crop full of Aphis-excretions shares it out with many of the nurses inside the nest. A worker may also solicit another worker for food, standing face to face and tapping or strok- ing the other with her antennae. Clearly the donation of food at least conveys the information that such food is available at some point outside the nest. Conversely, the eagerness with which food is demanded, depending on the state of the grubs as well as on that of the adults, conveys to the foraging workers information on the needs of the whole colony. But it has not yet been possible to detect the communication of precise information, either on the location of food or on the needs of the colony. In recent years, however, a good deal has been learnt of how ants find their way, and this bears on the method by which a worker stimulated to start foraging might successfully find a source of food already discovered by other workers. After all, even the marking of a trail involves a simple form of language.

TRAIL-MAKING AND PATH-FINDING

As might be expected, there are considerable differences between different kinds of ants : many of them, though rather versatile, tend to use one type of information more than another. Thus the shining jet black ant, Lasius fuliginosus, mainly follows a scent-trail, like a blood-hound. In this species well-beaten trails are formed, often visible to the naked eye since most movable rubbish has been cleared away. Down these trails rather dense, close columns of the foragers may often be seen marching. On the other hand, the common British black ant, Lasius niger, mostly uses its eyes to see the way and has much more indefinite trails, along which it usually goes singly. Most species, however, can when necessary use senses other than the favourite one. The black ant, for instance, certainly makes use of scent as well as of sight.

One method by which ants can find their way is known as the " sun compass reaction ", and was first studied in North Africa by Santschi about forty years ago. When using the sun as a clue to the direction, ants will continue to walk on a path parallel to the one leading to the nest even if they are picked up and put down some way off the track. They also distinguish between the way to and the way from the nest. This is probably accomplished by arranging that the sunlight falls on one eye so that the ants walk at a constant angle to it. In the short time necessary for making one journey the sun does not move enough to make the method inaccurate. But when an ant is confined for two or three hours in a dark box, placed over the track, the sun will move through an appreciable angle. As a result, when the ant is liberated it takes a course which makes about the same angle to what would have been the correct course at the time it was first confined.

Carthy, in London, has recently studied the way-finding of the jet black ant and the black ant in a small artificial arena. The glass floor of the arena can be turned round independently of the circular sides. Some grubs are placed in the centre and a worker is allowed to enter through a hole in the

sides, find the grubs, and start carrying them back to the nest.
While the ant is at the centre, it can be confined in a small
thimble and the arena rotated. Some species, such as the
black ant, are not much disturbed by this rotation, but
usually set off in the right direction. They probably mainly
use their eyes to find the way. The jet black ant, on the
other hand, is usually nonplussed. It tends to follow the old
scent-trail, and this, after the rotation, no longer leads to the
door in the wall. Carthy was able to make the scent tracks
visible by dusting them with very fine Lycopodium powder.
The track appears to be formed by liquid excreta which is
ejected when the abdomen is dragged over the floor. In
another ant, Myrmica, Macgregor showed that the track
consisted of a discontinuous series of pear-shaped spots.

Some of these ants are also sensitive to the plane of polarisation of light, like the honey bee. The light by which the
ant finds its way across the arena can be polarised by passing
it through a sheet of " polaroid ". If then, before the return
journey, the sheet of polaroid is rotated through an angle,
the ant is much less certain of the correct way back. There
seems little doubt that with further experiments much more
will be discovered about how ants find their way to and from
the nest. In so far as all the ants in one nest use the same
tracks, this already has some social significance, but any
discovery of definite system of intercommunication would
have much more.

The special feature of the social life of ants, compared
with that of bees or wasps, is the much wider range in
behaviour and in type of social organisation. This is partly
because there are many more kinds of ants than there are of
the other social Hymenoptera. This in turn is to some
extent a reflexion of the long period over which they have

gradually perfected their organisation. Some members of the earlier types of social species, such as the Ponerines, still exist, and the most highly organised types have existed long enough to have evolved in widely different directions. But it does seem also that many species of ant are less rigidly specialised in their behaviour than any wasp or bee. Many of them are able to use a variety of foods, and their apparently simple nests are really more adaptable to a varying environment. In this way the ants have become one of the dominant groups of insects in all the warmer parts of the world.

7

SOCIAL PARASITES

The long history of insect evolution has largely been a struggle for food and shelter. It has often evidently proved easier to make use of the energy and labour of another species than to be completely independent. Any habitat in which many animals live will be attractive to a large set of hangers-on which depend in various ways on the activities of other species. The more diversified the fauna, the more opportunities for still more species to behave as predators, parasites or guests of various kinds. The nests of social insects, providing excellent shelter and rich stores of food, have proved particularly attractive to other species. Moreover, the nursing instincts which become so highly developed in the most successful social species make them especially liable to deception. The nests of ants and termites, and to a less extent those of bees and wasps, provide a home for a large variety of strange insects. Many of these are merely scavengers and survive by being inconspicuous, by having impenetrable armour or by some type of chemical defence, such as a repellent glandular secretion.

ANT GUESTS

The most successful and remarkable of the guests have developed special secretions to which the social insects are powerfully attracted. In some instances the attraction may be regarded as dangerous socially as drug addiction in man.

A number of beetles, for instance, have tufts of yellow hairs which produce some aromatic substance to which ants are strongly attracted. The ants lick the hairs and take as much care of the beetles as they do of their own brood. Yet often the beetles feed on the ant-grubs and are dangerous parasites of the colony. The caterpillars of some of the blue butterflies have a gland which produces a sweet secretion which ants eagerly lick up. Some of the caterpillars are carried into the nest by the ants and carefully tended. And yet the food of the caterpillars is either the aphid ant-cows, or else the brood of the ants themselves. Other insects, like the flies observed by Farquharson in Africa, while not parasites of the colony, have evolved a way of soliciting a meal from the foraging workers so that they can obtain a share of the food intended for the grubs.

SOCIAL PARASITES

All these insects are parasites of social insects, but are not themselves social. But there is another group of parasites whose behaviour is more like that of the cuckoo who lays her eggs in the nests of other birds. These are the real social parasites : species of wasps, bees and ants, mostly without a worker caste of their own, which lay their eggs in the nest of some other species of their own group. This behaviour is evidently not so much a response to the direct struggle for food and shelter as to the difficulties and dangers of founding a new colony. Amongst the social wasps certain species both of Vespula and of Polistes have become social parasites of this type. The cuckoo Vespula, of which there are three in western Europe and one, V. austriaca, in England, can be recognised in the queen by slight differences in the shape of various parts of the head. The differences are not very pronounced, but have been supposed to be advantageous in fighting. There is little sign of loss of the tools used in nesting, such as occurs in cuckoo bees, though the shape of the mandibles is thought to be less suited to the manufacture of paper. The cuckoo wasps have no workers.

It is curious that V. austriaca is a short-headed wasp and lives with a short-headed host, V. rufa, whereas the two other cuckoo Vespula are long-headed wasps and live with long-headed hosts. Yet all three cuckoo species show similar changes in head-shape which must have been acquired independently as a result of their mode of life.

Our knowledge of how the cuckoo wasps establish themselves in the nest of the host is still very fragmentary. The most extensive observations are those of Weyrauch in western Germany. The cuckoo queens come out of hibernation a good deal later than their hosts, and attempt to enter the nest at about the time that the first comb of queen cells is being constructed. At some point, perhaps when the cuckoo begins to lay, a big fight ensues with the workers, many of which are found dead in or near nests containing a cuckoo queen. The host queen may survive for a time, but apparently is killed or dies before long. Eventually, the combs produce some workers and a few males of the host, but no young host queens. Most of the later brood consist of young queens and males of the cuckoo wasp.

Amongst the wasps of the genus Polistes also, there are three cuckoo species in western Europe. Their existence was hardly suspected until it was first established by Weyrauch in 1937. More detailed studies were published in Switzerland by de Beaumont, in 1945. All the species have a characteristic groove on the mandibles, due to a thickening of the upper and lower margins. This, together with certain other features of the head, are thought to be useful in fighting, but the cuckoo Polistes are less different from the industrious species than are the cuckoo Vespula. It is not even certain whether one of them may not produce a few workers, for in this group it is difficult to separate queens from workers unless the wasps are dissected.

The way in which the cuckoo queen establishes herself is not known, but it seems that the host queen is somehow eliminated, perhaps after a fight. The brood which is later produced consists mainly of males and females of the cuckoo species, with a few workers and males but no queens of the host. Even the later host workers hatch only from cells occupied before the cuckoo queen takes over. It seems significant that in both these kinds of cuckoo wasp it is the young queens of the host brood that are particularly eliminated. This is not to the advantage of the parasite, since it means that there will be fewer nests to invade in the following year. It suggests rather that queens are, so to speak, "expensive" for the colony to produce. It would then be difficult either from the point of view of labour or of the supply of some scarce food to produce queens of two kinds.

The cuckoo humble bees of the queens Psithyrus are more different from their hosts than are the cuckoo wasps. The name of these bees means "whisperer", because the female in flying makes a much less pronounced hum than does a true Bombus ; but the most striking difference in this sex is the loss of the pollen-collecting apparatus. There is no pollen-basket on the hind leg and instead of a shining, rather concave surface there is a convex, densely hairy one. The sting is more powerful, and the whole skeleton is stronger and more thoroughly articulated against attack. The worker caste is absent.

As in the ordinary humble bees, males and females are produced in the autumn, and after fertilisation the females hibernate in a shallow hole in a bank. When they come out rather later in the spring, instead of beginning a nest of their own they search for a nest of a humble bee of the correct species, since each kind of parasite is attached to one or a few host species. The entry into a nest is a dangerous process, and is often unsuccessful if the host colony has reached a moderate size. The smaller the colony, the more easily it is invaded. The resistance is due entirely to the workers, and if there are enough of them the Psithyrus queen may be killed. One or more Psithyrus may be found dead in the tunnel leading to a humble-bee nest. If an entry is successfully accomplished, probably after a few of the workers have been killed, the cuckoo queen becomes established in the nest and leaves it no more. For some time she frequently mauls workers when they meet inside the nest, usually without killing them. Eventually an uneasy truce is established.

There has been disagreement about the relations between the cuckoo humble bees and the host queens. Sladen, who published in 1912 a well-known work on British humble bees, maintained the host queen was always killed. Hoffer in Austria and Plath in the United States, however, have numerous records of the two queens living side by side and taking little notice of one another. Cumber also found a nest in which both the Bombus and its parasite had been laying eggs. Some of these discrepancies may arise from the different behaviour of different species, but it seems almost certain that Sladen exaggerated the antagonism between the queens. Nevertheless, once the colony has been invaded the Bombus never produces any young queens, very rarely any males, and probably no more workers. This would be quite understandable if the Bombus queen were killed, but it is not clear how it happens when she is not. Possibly the cuckoo queen eats the eggs; or it may be that, as described below in some of the ants, the behaviour of the workers becomes perverted; the colony therefore produces a brood of Psithyrus males and queens and then dies out rather earlier than usual, since the host workers are not being replaced.

HOW CUCKOO BEES AND WASPS AROSE

It is possible to form some idea how this type of behaviour may have arisen in bees and wasps, and this applies also to the solitary kinds of these insects, amongst which there are quite a number of cuckoo species. In both wasps (Vespula) and in humble bees, suitable nesting sites are not very numerous, and this seems often to set a limit to the abundance of a species in any locality. As has been stated earlier, this leads to a certain amount of fighting even between the ordinary industrious species. Queens which have tried to force their way into an already occupied site are quite commonly found dead at the entrance. Sladen showed that in two very similar British humble bees, B. terrestris and B. lucorum, the former sometimes succeeds in invading young nests of the latter and taking them over. Thus, after a short period in which there are workers of both species present, the B. lucorum workers die out because they are not being replenished, and the colony becomes a pure one of B. terrestris. This behaviour, though exceptional for these species, is very similar to the " temporary social parasitism " of certain ants, mentioned in the previous chapter. Such successful invasions by normally industrious species seem to be very rare in wasps, but Nixon in 1935 recorded one example: a nest of Vespula vulgaris taken over by a queen of V. germanica.

In 1927 I showed that cuckoo bees, and in particular the species of Psithyrus, tend to have the colours and shorter hairs which are often seen in southern races of industrious species. Bischoff and Weyrauch have noted that cuckoo wasps also have a "southern" appearance. It has been suggested that if a southern species works its way into more northern territory already occupied by some allied species, the situation would be one in which cuckoo habits might well arise. The more southern species would, as is known from the observation of industrious species, tend to leave hibernation later in the spring. The conditions for founding a colony might well be more unfavourable than in southern regions. At the same time, the northern species, having come out earlier, would already have occupied many of the suitable nesting places.

This is in fact what seems to happen with Bombus terrestris and B. lucorum. In Europe as a whole, terrestris has a more southern distribution than lucorum and where, as in England, they occur side by side, terrestris tends to come out of hibernation later. This probably explains the occasional invasion of nests of lucorum by terrestris, and it is possible that over a long period (perhaps to be measured in millions of years) such a conjunction of species might lead the southern intruder to become a cuckoo species.

This hypothesis may account for the evolution of many of the cuckoo wasps and bees, both social and solitary, in Europe and North America. It can hardly account for the origin of the numerous solitary cuckoo species in the tropics. So far the only known social cuckoo species of bee or wasp outside the temperate regions are two Psithyrus found in eastern Asia. Possibly in tropical climates cuckoo habits might tend to develop quite apart from climate whenever one species invaded the territory occupied by one of similar habits. However, much more observation of the tropical species is required before such speculation can have much value.

PARASITIC ANTS

The story of the development of cuckoo habits in ants is more complicated. It has happened in a much bigger range of species, and may have arisen in more than one way. Moreover, all sorts of stages in the development of such behaviour are known, ending up in workerless, degenerate parasites.

One manifestation of such behaviour is the temporary social parasitism of such species as the wood ant, Formica rufa, mentioned in the previous chapter. A rather different example is seen in the North African ant, Bothriomyrmex decapitans. The queen of this species, after her marriage flight, enters the nest of another ant, Tapinoma nigerrimum. She does this by alighting near the nest and allowing herself to be dragged in by the workers. She is subjected to some rough treatment, but once inside the nest, she takes refuge on either the head or the back of the much larger Tapinoma queen. In either of these positions she seems to be safe from the workers. After a time, she bites off the head of the Tapinoma queen, but since by that time she has (presumably) acquired the characteristic odour of the nest, she is readily adopted as the new queen and her brood is reared. Gradually, all the Tapinoma workers die off, and a pure colony of the Bothriomyrmex remains.

Similar and, from a human point of view, even less pleasant behaviour has been observed by Bruch in an Argentinian ant, Labauchena. Several of the very small queens of this species invade a nest of the fire ant, Solenopsis, and climb onto the back of the much bigger queen. There they co-operate in gnawing off her head, an act which may take them many days.

Such methods of colony foundation are appropriate to

species with small queens which could not found colonies unaided. They can scarcely be considered degenerate, though they may have led to the slave-making habit of such species as the blood-red ant, Formica sanguinea. When the young queen of this ant founds a new colony, she invades a small nest, usually of Formica fusca, and kills the queen and any of the workers which resist or attack her. Before long, all the adults of the host ant will be dead and the sanguinea queen tends the remaining brood. When new young workers of fusca emerge they look after the brood which hatched from the eggs laid by the sanguinea queen. In some regions, and in some varieties of the blood-red ant, there is no further development of parasitism, and as the fusca workers die out, the colony becomes pure sanguinea. This behaviour is no more than a form of temporary social parasitism in which the relations between the host and the invader are even less cordial than usual.

More commonly, particularly in western Europe, including Britain, the later history of the colony is different, since the population of slave workers is maintained by a regular series of raids on other nests. These are carried out by sanguinea workers which leave the nest in a large straggling group which may cover several yards. Sometimes several smaller groups follow one another at short intervals. When this happens the first to arrive surround the nest of the prospective victims. When the whole group of raiders has assembled they enter the nest and begin seizing grubs and cocoons. Workers which resist are killed, but the others are not molested. Eventually, each sanguinea worker returns, carrying a grub or cocoon.

This remarkable behaviour attracted the attention of Darwin, and in *The Origin of Species* he describes a number of his own observations on this ant. Darwin was puzzled how the habit of raiding could be transmitted from generation to generation if, as he supposed, it was manifested only by the sterile workers. Much later, however, it was discovered that the behaviour of the queen in founding new colonies is not so very different from that of the workers. Moreover, as Darwin remarks, the queens whose worker offspring behaved in a way which made the colony as a whole most successful would be the ones which leave most offspring. That is, the process of natural selection involves evolutionary change in the whole colony, rather than in single individuals. The sterility of the workers from this point of view is immaterial.

Darwin also suggested that the raids are partly for obtaining food, like the raids of driver ants, and that the slaves hatch from grubs or cocoons which fail to get eaten. There is little doubt that this theory is partly true. Sanguinea does eat a varying proportion of the grubs and cocoons which it brings home, and does sometimes raid nests of species which it never uses as slaves. Indeed, some other species, such as the wood ant, whose queens found colonies as temporary social parasites, do also raid other ants' nests for food, at least occasionally.

It may be that the slave-making habits of sanguinea have developed from two directions; partly from the way in which the queen founds new colonies and partly from food raids undertaken by the workers. One remarkable feature of the raids is that they may be made on a nest as distant as 100 yards from the sanguinea nest, and that on the way to the chosen nest the ants may pass and ignore other nests of the slave-providing species. This suggests that scouts have previously reconnoitred the way. The ants are not, however, led by the scouts, since the position at the head of the raiders is occupied by new individuals at frequent intervals.

Although most of the nursing in the sanguinea nest is done by the slaves, at least in those colonies which have them, the species is not at all degenerate. The workers are fully capable of running the whole colony if necessary; they are extremely active and, judging by their behaviour on raids, very well endowed. Nevertheless, it is probable that the behaviour of some of the more degenerate parasites described in the next few paragraphs developed out of slave-raiding. Once the ability to run the colony independently has been lost it is evidently very difficult to regain it.

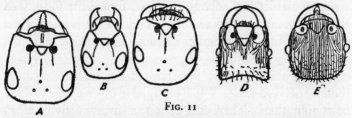

FIG. 11

Heads of some workers of ants and of their slavemakers or parasites. A, the slave-maker, Formica sanguinea; B, its slave, Formica fusca; C, the amazon ant, Polyergus rufescens; D, Strongylognathus, parasite; E, Tetramorium, host. The mandibles of types C and D are suitable only for offence; such ants have to be fed by their slaves or hosts.

This is seen in the more specialised slave-keeping ants known as the amazons, the various species of Polyergus. Some of these are found in western Europe (France, Switzerland), others all the way across central Asia to Japan and again in North America.

The mandibles of the amazons are sickle-shaped and pointed, without the broad, finely-toothed edge of those of an ordinary ant. These jaws are unsuited for anything but fighting, and the amazons are incapable of making a nest or of looking after their own young. The tongue-like organ of the mouthparts is also short, and they are almost incapable of feeding themselves. At the most they can take a little liquid food if they happen to find a conveniently placed drop. The amazons are thus clearly degenerate, and can live for only a short time in the absence of their slaves. There are few things more remarkable in an amazon colony than the sight of large groups of workers, much bigger and more formidable than their slaves, spending the whole day doing nothing but occasionally cleaning themselves or soliciting the slaves for food.

The small slaves, which are usually workers of Formica fusca, construct and enlarge the nest, forage for food, feed the amazons and tend all the brood. This is in marked contrast to Formica sanguinea, whose slaves never leave the nest but only tend the brood.

During the height of the summer the amazons make a number of slave-raids. When engaged in this way their whole behaviour alters. There is a period of intense activity when the amazons gather round the mouth of the nest, and then they all start off in a rather compact army. Normally they go straight to some nest of the slave species, and it seems that, as in Formica sanguinea, the way must be surveyed in preparation, though the army is not led by any particular individual. When they arrive at the slave nest they pour in through the entrance and begin to come out with grubs or cocoons. Adult ants are not attacked unless they resist, when they are killed instantly by being pierced with the sickle-like mandibles. The return with prey is more leisurely and not made in a compact group. Much of the prey serves later as food for the colony, but some of the cocoons are kept to produce new slaves.

When the amazon queen founds a new colony, after her marriage flight, she enters a colony of Formica fusca or one of the similar species. She kills the queen but does not molest the workers ; these will later feed her and tend her brood. The workers which she has taken over are afterwards replaced by new slaves obtained by the raids of amazon workers, so that the colony never becomes pure for one species as in a temporary social parasite.

PARASITISM AND DEGENERATION

There are a great many more kinds of parasitic ants, and they all have sickle-shaped mandibles and are in various degrees incapable of maintaining the colony unaided. A few seem to be slave-keepers rather after the pattern of the amazons, with pugnacious workers. But many are more degenerate and occupy a relatively subordinate position in the host colony both as regards numbers and behaviour. They have, in fact, become much more like true parasites, and their chief effect in this case is to prevent normal reproduction by the host species. All the more degenerate parasitic ants have lost the worker caste. Many of them are found in only a few localities and are rarely seen. It appears that such behaviour is not ultimately a reliable way of perpetuating the species. Moreover, once a species has reached the stage realised in the amazons, there is no return.

In Great Britain there are three species of ant which are permanent social parasites and which, though not closely related to one another, are examples of three stages in degeneration. Formicoxenus nitidulus lives exclusively in the nests of the very much larger wood ant, Formica rufa, and some other very similar species. It excavates small chambers in the recesses of the nest and looks after its own brood. Owing to its small size the galleries it uses are inaccessible to the workers of its much larger host, but it is able to wander freely throughout the Formica nest. Its natural food is unknown, but it will eat honey or ant grubs in observation nests. It does not seem to attack the Formica brood, and the two sorts of ants take very little notice of one another. It is really dependent on the host only for shelter and protection, and is thus hardly a parasite. Structurally, it shows few signs of degeneration, since workers are numerous and have normal broad mandibles. The male, however, is wingless and very worker-like in appearance. There can therefore be no marriage flight, and mating takes place on top of the Formica nest.

The virgin queens are winged, and presumably the majority of them fly off to found new colonies, after mating with their wingless brothers. New colonies are founded in small cells excavated on the edge of a Formica nest. The queen and her young brood must move into the interior at a later stage. In full established colonies, if the Formica moves its nest to a new site the Formicoxenus follow them, carrying their own brood.

The various species of Strongylognathus are essentially slave-makers, more dependent than the blood-red ant but somewhat less so than the amazons. They have the characteristic sickle-shaped mandibles of all the true parasites. One species, S. huberi, which is found in Switzerland, is known to make raids to obtain the brood of the ant Tetramorium caespitum. The raids are apparently made at night when the Tetramorium are less active and less resistant to aggression. The case is unusual in that some Tetramorium slaves may take part in the raids, but few observations on the subject have so far been made. While the Tetramorium slaves are necessary for rearing the brood, the Strongylognathus may do some of the work of nest-construction. Another species, S. testaceus, is quite common in parts of western Europe and is more degenerate. The worker caste is never numerous and seems to be on the way to disappearance. It does not make slave-raids, but the young queen invades a Tetra-morium nest and lives there without killing the host queen. Both queens continue to lay eggs, but those of the host queen produce thereafter nothing but workers, whereas the brood of the parasite consists of all three castes but especially of males and queens. The species is thus harmful to its host—since it prevents normal reproduction—and degenerate, because it is unable to rear its own brood. This is done for it by the Tetramorium workers. In 1935 a new species of Strongylognathus, S. diveri, was discovered by Diver, in Dorset. It is very similar to S. testaceus, and may be a form of that species. It certainly lives in the nests of Tetramorium, and it has few and rather feeble workers.

The third British parasitic ant is Anergates atratulus. It too lives in the nests of Tetramorium. In England it has been found only in the New Forest, and there only on very few occasions. It is also rare on the continent. It is an example of a group of rare species, each belonging to a different genus, and probably showing the last stage in a long parasitic history. Such forms are rare, it is thought, because they have passed the point where parasitism is even a temporary advantage and are now approaching extinction.

Anergates has no workers. The male is wingless, of a pale colour, and with a rather soft skeleton, more like a pupa than an adult insect. The virgin queens are relatively small; after mating with their brothers in the colony which pro-duced them, they fly off to try to invade new colonies. If successful, they kill the host queen and begin laying eggs at a great rate, their abdomen swelling to an enormous size. The host workers care for the brood, but the whole colony must die out rather quickly since the workers are no longer being replaced. Although the species can live only in places where Tetramorium colonies are numerous, it tends to reduce their number by its parasitic action. The life of the species is therefore rather precarious.

SUMMING UP PARASITISM

The story of the parasitic wasps, bees and ants which avoid some of the difficulties of colony foundation by their dependence on other species is suggestive in several ways. There is of course no need to draw the moral that—even in organic evolution—crime does not pay. Many parasites, both amongst the insects and in other animal groups, are highly successful and show no signs of dying out. Some dependence on other organisms is universal amongst animals, for all depend ultimately on the green plants. A successful parasite, however, is one which does least damage to its host, which allows if not the individual at least the species to flourish. Apart from Formicoxenus, all the types dealt with in this chapter prevent their host from producing sexual forms. This by itself might not be a fatal objection to their mode of life, from the evolutionary point of view. Yet once the parasitic habit has begun, there seems to be a general tendency for the worker caste to be lost.

This loss can be looked at in two ways. First, in any species, varieties lacking structures or types of behaviour which contribute towards successful reproduction and survival will tend to be eliminated. In most social insects, the presence of many workers is essential to successful reproduction in the colony ; any tendency to reduce the number of workers would be suppressed. But in a parasitic species whose brood may be reared by the host workers, the workers of the parasite are no longer essential to its survival. A situation therefore exists in which the suppression of the worker caste will not be an immediate disadvantage.

Secondly, while we are uncertain what determines whether

a particular egg produces a queen or a worker, it seems that the balance between the two possible types of development may be quite easily tipped one way or the other. It may well be that the original balance of the colony is upset by the presence of the parasite in such a way as to suppress queen-production in the host. It has been suggested, for instance, that in Strongylognathus it requires less effort to rear the small sexual forms of the parasite than the much larger ones of the host, Tetramorium. This might lead both to the suppression of the sexual forms of the host and to the conversion of the majority of the parasite brood into sexual forms rather than workers. This argument, of course, raises again the question whether castes are determined by feeding or genetically.

Probably in the more degenerate parasites such as Anergates the power to produce a worker caste has been completely lost as the result of a long process of selection. Once the workers have lost almost all their functions it is more advantageous to have all the females fully fertile. It is possible that Strongylognathus provides an example of what has been called social regulation, a process (as we shall see) much better known in the termites. This is the method by which different types of individuals and of developmental stages are produced in the proportions best suited to the needs of the colony. It arises from the natural interplay of the normal behaviour of the species with its environment. In Anergates, social regulation has been replaced by an innate physiological process.

Social parasites are not very numerous, perhaps because species which have evolved along these lines are rarely successful for long. Amongst the eighty-eight social species in Great Britain, eleven are parasitic, if one counts the slave-making ant. According to our present knowledge, the proportion is lower in tropical regions, and it is possible that the difficulties of colony foundation in a temperate climate have favoured the development of such parasites.

It is a curious point that slavery and parasitism of this type do not seem to occur in the termites. A few cases of compound nests, rather like those of Formicoxenus and Formica, are the nearest approach to it. It is not clear why termites have not shown this type of evolution. It may be partly because in them the queen is always attended by a male and mating has often to be repeated. The invasion of a strange nest by the two sexes together might be more difficult. Another possibility is that it would be harder to eliminate the reproduction of the host species, since when the original queen is killed alternative queens can readily be produced. A similar argument might be applied to the honey bee and its allies, since these are not subject to attacks by cuckoo species. Such suggestions, however, are at the best very tentative, and there is still much in the evolution of the parasites which we do not understand.

8

THE TERMITES

The termites, sometimes called " white ants ", are the least familiar of the social insects to Europeans. Although one or two species are found in France and Italy and fifty-five in the United States, the group is essentially tropical, and even in warmer lands the insects are usually inconspicuous apart from some of their large nests. There are some remarkable analogies with other social insects, and since the majority of individuals are wingless, it is with the ants that they would naturally be compared. Yet in reality the differences between the termites and all other social insects are profound and more important than the resemblances. It is almost true to say that what they have in common is little more than the minimum which allows us to apply the one word " social " to both groups.

The peculiarities of the termites are traceable to two basic facts about their societies. First, males and females are represented at all stages and in all castes, and this is associated with a different method of sex-determination. Secondly, development is gradual, over a series of moults, with no helpless grub-stage. At least after a few moults, each individual is a potential worker and its bodily form and functions can partly be determined by the number of moults it has experienced. Thus the whole caste system is much more flexible and complicated. In the ants, apart from the exceptional forms which use the grub to spin silk, the whole work of the colony is done by adults, and by definition the form and behaviour of these is in the main fixed for life. Though there may be variation in fertility and changes in the size of the abdomen, as in the queen legionary ant or the worker honey-pot ant, the form of most individuals is constant and none can alter parts of the body other than the abdomen. In the termites, on the other hand, the individual is usually capable of changing both its appearance and its functions over a considerable period after it has begun to play an important part in the life of the colony. This introduces the possibility of "social regulation", the moulding of the individual to the needs of the community at any particular moment.

NESTS

Termite nests often form a conspicuous feature of the landscape in tropical regions. They vary much in shape and structure, even within one group. They are built of earth, of rotten wood, or of the insects' own excrement, or of a mixture of these. Often a sort of cement is made of saliva and clay, and this may set extremely hard. One of the difficulties of preparing the site at Kongwa in East Africa for planting ground nuts was the problem of removing the numerous large termite nests. Often a nest required a large charge of dynamite.

Termites, apart from a few exceptional species, avoid all contact with light and with the free air. Their nests are therefore always closed. When the winged forms wish to leave, or in species which send foraging columns into the open, temporary exit-holes are made, but they are soon closed again. A few species are known to carry their excrement outside the nest and these also make temporary exits for the purpose. More commonly, if the excrement is not used in building, it is piled in certain chambers in the periphery of the nest.

The most conspicuous type of nest is a tall, steeple-like structure which may be 10–20 feet high and up to 50 feet in circumference at the base. The outer walls of these nests are as hard as concrete. Inside, the nests are usually divided up into a number of chambers of varying size and shape, connected by passages. The more central of these chambers are inhabited by the royal couple—king and queen—by the young brood and by various stores of food. Such nests may also extend several feet underground. In other species there is a central chamber suspended inside the much larger outer part of the nest, almost like a wasps' nest hanging inside a hollow tree.

Some of the large Australian nests have the steeple-like structure very flattened, so that it is almost tongue-shaped. These are known as " compass termites " because the two broader faces always point east and west. Although very useful to the bushrangers, the value of this arrangement to the termites is not known.

Another type of conspicuous nest is built on trees. These may be of irregular shape and placed flat against the trunk or in a fork of the branches. Others are spherical and are built all round thin branches. These may be conspicuous objects in tropical rain forest and have been known as " niggers' heads ". The nests on tree trunks are sometimes surmounted by a number of V-shaped ridges built out from the trunk which seem to protect the nest from water running down the trunk. Another construction which has the same effect is a series of overlapping flaps, shaped almost like a hand with fingers and projecting all round one side of the tree-trunk. Nests with such protections against rain are built by various species in the damp forests of Africa and South America.

The nests so far described are the conspicuous ones, but many kinds, probably the majority, are hidden. Some, especially perhaps those in drier regions, are entirely subterranean. These may be merely a series of irregular interconnected underground chambers or they may have a much more definite structure, an enclosed, walled-in underground space within which the actual chambers are constructed. Some of the desert-living species make vertical tunnels many feet long to reach the deep-lying water table so that they can bring up water to keep the nest suitably damp.

The two groups of termites which have been most intensively studied (because they happen to be represented in Europe and North America) make very simple nests in decayed wood. Irregular branching tunnels are excavated, sometimes widening into chambers ; but there is no special provision for the royal couple or the brood : these merely occupy the more central part of the colony.

THE NUPTIALS

Just as in ants, new colonies of termites normally arise from flights of winged sexual forms. Commonly, these forms are produced only during a short season when they may appear in millions. In parts of Africa, when camping in the bush, it may be difficult to eat one's supper by lamplight because of the immense numbers of winged termites which are attracted to the light and fall into the food. Other species, however, fly by day.

Some time before the marriage flight the workers construct special passages to the outside, each wide enough to allow several winged adults to leave simultaneously. One species is known to construct special waiting-rooms for the winged forms to stand in, near the exit-holes. At the moment of departure some of the workers and soldiers, which for some time have shown signs of excitement, pass out of the exit-holes and form a ring round them, outside the nest. The winged forms then leave and fly off over a period of two or two-and-a-half hours; thereafter the holes are stopped up again.

Few termites are capable of powerful flight. Usually they travel less than 100 yards, only occasionally as much as a mile. Mating does not usually take place in the air, though in Pseudacanthotermes the male seizes the female in flight, losing his wings at the same moment. The female flies on with the male and loses her wings only on landing. In most species, however, both sexes get rid of their wings on landing, pressing them against some plant stems so that they snap off at the preformed line of weakness.

The wingless adults are still very excited and dash about in all directions on the ground. If a male and female meet, they stand face to face, tapping one another with their mouth-parts and antennae. If the male is accepted, the female turns round and walks off, followed by the male, and they undertake what is known as " the nuptial promenade ". This is essentially a search by the female, often lasting several hours, for a suitable site for a new nest. This will very often be found at the side of a fallen log, or against a post or tree, preferably where conditions are a little damp.

The male and female together dig a tunnel, later enlarged into a small chamber. The entrance is gradually blocked by debris from inside, the complete work taking a day or two. When they have finished they become inactive, and at this moment, for some obscure reason, they each remove the last few segments of their antennae. Mating occurs for the first time either at once or some time up to a fortnight later, according to the species. It can be well imagined that flights of winged termites cause great excitement in all the beasts of the field and birds of the air, and that enormous numbers of the termites are eaten before some are able to dig their nuptial chamber. In South America nests of social wasps are often found to be stocked with winged termites, and some South American Indians are fond of eating them.

FIG. 12

Scene in the royal chamber of the African Termes bellicosus showing the king, queen and attendant soldiers and workers (after Escherich).

THE COLONY AND ITS FOOD AND AGRICULTURE

In the less specialised termites the young king and queen eat rotten wood during the early part of colony-foundation ; from it they elaborate food which they supply to the first stages of their brood. In them, too, the brood quite soon are themselves able to feed on wood and later to take over the supply of food for the royal couple, who eventually no longer feed themselves at all. In the most specialised termites the nuptial chamber is often made in the soil where there is no external source of food. The first brood is brought up entirely by the royal couple, who rely on what was in their intestines at the time of flight, on the enormous fat-reserves which have been built up inside their bodies, and on material derived from the degenerating wing-muscles. In these termites, too, the brood do not become self-supporting until they are much bigger. In some species it is perhaps only the fully grown workers which actually obtain new food, and all the other inhabitants have to be fed by them.

Termites are pre-eminently eaters of wood; to a less extent they eat other vegetable matter. The less specialised types are almost exclusively wood-feeders, and these also nest in timber. More highly organised forms make use of other supplies, such as dead vegetable matter in general. A few collect grass and other green leaves which they cut like leaf-cutter ants. Others collect grass seeds which they store in underground chambers after the manner of harvesting ants.

Many, especially those which make the large conspicuous nests, store varied vegetable material in underground chambers. These collections, especially those of thoroughly chewed wood, become impregnated with fungus just as happens in leaf-cutting ants, and some of the fungi are peculiar to this situation. The termites feed on the fungus, but it does not seem that they are so closely adapted to the diet as the ants are. Almost certainly it is one of the ways in which they can supplement their diet with proteins or with special growth-promoting substances. Termites also eat all the moult skins of the developing brood, having a digestive juice which can dissolve insect skeletons. Any wounded or damaged individual is quickly eaten by the others. A few species are known to collect other termites and to store them in chambers just as they do vegetable materials. Like the queens of other social insects, the termite royal couple may eat an appreciable number of their own eggs, even when there seems to be no shortage of food.

The termites which are more especially wood feeders have a remarkable digestive specialisation. Their intestine contains vast numbers of protozoa belonging to a special group which is found only in this situation. These microscopic unicellular animals are able to digest the cellulose which forms the major part of all woody tissues. They convert the cellulose into sugars or into other substances which the termites can assimilate. Although wood might provide a complete diet, it would be an extremely wasteful one if none of the cellulose could be absorbed. The relation between termites and their intestinal fauna is a typical example of a symbiosis: the protozoa receive chewed wood in a suitable environment, while the termites receive back digested cellulose and also assimilate a fair proportion of the protozoa themselves.

The young brood quickly become infected with the protozoa by eating the faeces of older members of the colony, and the royal couple carry some infected material with them when they go on their marriage flight. The process has an interesting analogy with the digestion of such animals as the cow, in which the rumen, a special compartment of the intestines, contains vast numbers of bacteria which help to digest grass or hay.

About twenty-five years ago, Cleveland showed that the protozoa in the termite intestine could be eliminated by exposing the insects to a raised pressure of oxygen. After this has been done the termites can no longer assimilate cellulose, and, if given no food but purified paper, die rather rapidly. Conversely, normal untreated termites can subsist on such a diet for a long time. Those termite species which seem to have a more varied natural diet lack the intestinal fauna; exactly what fraction of their diet they assimilate and how they do it is unknown.

Much of the food which termites eat does not come directly from outside the nest but has already been partly assimilated or even secreted by other members of the colony. As already noted, in some species it is only the workers which feed on unprepared foods, such as wood-fragments. Treated food is of three kinds. Thoroughly chewed and probably partially digested material may be received from the mouth. This source also provides saliva, which is the richest and most thoroughly prepared food. Finally, termites can pass two types of faeces ; one is relatively dry and is thoroughly digested, waste material. It is thrown away or built into the nest. The other consists of much more liquid material which is really food which has passed through the intestine without having all nutritive substances extracted from it. It contains large numbers of the intestinal protozoa in those termites which are infected. Generally, in a well-established colony, the youngest brood are given only prepared food, and soldiers are incapable of feeding themselves. The royal couple, particularly the queen, are fed continuously on saliva. At the other end of the queen an almost continuous stream of eggs and also of certain liquid excretions is continuously removed by bands of workers.

Termites are continually exchanging food with one another and also licking one another's bodies, probably to ensure cleanliness. It is supposed that the mutual interest which such behaviour engenders is an important link in the social organisation.

TERMITES AS PESTS

The wood-feeding habits of termites are the cause of very serious losses, chiefly in tropical countries. In Queensland Mastotermes makes it impossible to construct sheep-pens with wooden posts. In many tropical countries the wooden parts of houses have to be protected, especially where they come in contact with the earth. Even the use of concrete pillars to support the ground floor is not infallible, since the termites will build covered runways up the concrete. The damage is particularly insidious since it is entirely internal : the first sign of trouble is apt to be the collapse of a beam. During the Second World War crates of military stores which had to be piled on the ground in such countries as New Guinea were often attacked. A few weeks later, when the crates were moved, they fell to pieces.

During his travels at the end of the eighteenth century, Humboldt noted how rare it was to find old books in South America, and this was attributed to the attacks of termites. In such countries, to-day, all books are treated with a solution of a poison such as corrosive sublimate, to protect them against termites and other insects. The protection of houses is more difficult, but special modes of construction, employing concrete, have been devised, and there are expensive ways of impregnating wood with preservatives.

A curious but less important source of trouble occurs when runways are constructed for African aerodromes. It may be very expensive to level out the hard, projecting termites' nests and the strips may require repeated treatment because the nests are continually being reconstructed.

THE IMPORTANCE OF TERMITES TO TROPICAL PLANT-LIFE

Termites play an important part in the economy of tropical nature in other ways which impinge on human interests, though less directly. Trapnell has shown that where they are really numerous, as in parts of Rhodesia, they greatly influence the type of vegetation. Just as in England the vegetation of chalk downs differs from that of sand or clay, so in Africa termite country may have a characteristic flora. This will indirectly affect all the other animals, and will also determine what crops can be most successfully grown.

The natives are well aware that the soil formed from broken termites' nests is unusually fertile. To some extent these insects do the work which in temperate countries is done by earthworms. Worms drag dead leaves underground and eat them, and incorporate dead vegetable matter into the humus in a form which is suitable for plant growth. Termites do much the same thing, and may be very important in dry climates in which ordinary decay would be rather slow.

CASTES

So far the different individuals of which the colony is made up have been only incidentally mentioned. The subject is much more complicated than in other social insects, but the nature of the castes and the factors determining their development lie at the root of their whole social organisation. There is a larger number of castes than in other social insects and also a bigger difference between the simpler and most advanced types.

The eggs are laid singly except in the primitive Australian genus Mastotermes; in the latter they are laid in groups of sixteen to twenty-four, stuck together in two rows. The appearance of this egg-group is rather like the similar groups laid by cockroaches. The eggs develop slowly, and even at quite high temperatures (20–$25°$ C) may take three to twelve weeks to hatch, according to species and circumstances. In the less specialised termites the rate of egg-laying is not very high, perhaps two hundred to three hundred a year, but in the higher forms extraordinary figures have been recorded. Emerson kept queens of some of the South American species and actually observed that 1,600–7,500 eggs were laid in twenty-four hours. One of the species which builds the large steeple-like nests in Africa has been seen to lay 36,000 eggs in twenty-four hours ; this corresponds to thirteen million a year.

The method of sex-determination is different from that of ants, bees and wasps. If the developing sexual cells are stained and examined under a microscope, it is found that those of the female contain an even number of heavily stained, rod-like bodies or chromosomes. The corresponding cells in the male have one less chromosome. Thus as far as sex is concerned, the hereditary constitution of the queen termite may be represented by the symbols AAXX and of the male by AAX. The unfertilised eggs all have the constitution AX, but the sperm may be A or AX. Thus when the eggs are fertilised, females (AAXX) and males (AAX) are produced in equal numbers.

Termites are one of the groups of insects in which development up to the adult stage is gradual. This means that there is no grub, nor an immobile pupa enclosed in a cocoon. The individual which hatches from the egg is like the adult except that it is small, has no traces of wings, and is sexually immature.

It is convenient to use the French terminology for referring to the immature stages, though it is not universally accepted. The form which hatches from the egg is the larva, and it continues to be one in spite of moulting, as long as it has no traces of wings. The nymph resembles the larva but has visible wing-pads. To produce a fully winged adult from the nymph will require several moults, at each of which the wing-pads will get a little longer.

The soldier is typically without wing-pads, is relatively

large, is at least partly pigmented, has the eyes small or absent, and has a modified head. The most usual type of soldier has large mandibles, rather like the typical ant soldier, but there is another type in some termites known as the nasute. This type of soldier has reduced jaws, but has the top of the head produced into a long snout, at the tip of which is an orifice. This is the opening of a gland which produces a copious, sticky, often white, secretion. They can use this to gum up their enemies, principally ants. While there is some doubt whether, in certain species, the soldiers are much use in defending the colony, this does seem to be their function in most. It is very striking in the species which make daylight forays above ground that the immense columns of workers are flanked on each side by rows of soldiers; this strongly supports the idea that their function is protective.

Soldiers are incapable of feeding themselves and do none of the work of the colony. They might be compared to the workers of the amazon ants except that they do not make slave-raids. In some of the more specialised termites there are two sorts of soldiers, quite distinct in size, but mandibulate and nasute soldiers never occur in the same species.

In the less specialised termites there is no definite worker caste and the main work of the colony is performed by the nymphs and the older larvae. There may, however, be a group known as false workers whose form is semi-stabilised though under certain conditions they may recommence growing and moulting. If they do so, they become either larger false workers or else the substitute sexual forms described in the next paragraph. In the specialised termites, a true worker caste is found and may, as in the soldiers, be of two distinct sizes. Members of this caste resemble large larvae, having no vestiges of wings, and no or only minute eyes, but they have a fixed form and are no longer capable of growth and moulting.

The substitute sexual forms are individuals capable of reproduction, though not winged and not leaving the nest for a marriage flight. They appear whenever one or both of the royal couple die or is removed from the colony. In bodily form they resemble larvae or nymphs, depending on their mode of production, but the wing pads are often smaller than in the average nymphs. Once a termite has become a soldier, it can no longer become a substitute sexual form, and the same holds for the true workers of the higher termites.

Grassé and his colleagues in France have obtained the conclusive experimental result in some of the lower termites that an artificial group of, say, twenty or thirty identical third stage larvae can differentiate in isolation into a colony in which all castes are represented. This self-differentiation of a group is the fundamental fact in caste-production. The higher termites, with their more elaborate caste system, have been much less studied, and little can be said about them with certainty. There is, however, no reason at the moment to suppose that their development is essentially different.

Another element of great importance in the flexible caste system is the influence upon it of the age and size of the colony. In general, the younger and smaller the colony, the fewer moults are necessary to produce the adults of any caste. In Zootermopsis, for instance, a soldier will be produced after five, six, seven or eight moults according as the colony consists of about twenty-five, fifty, one hundred or five hundred individuals, corresponding to ages of one, two, two to three, and four to five years, respectively. In such termites the winged sexual adults are not produced until the colony is several years old. This last point is another example of the natural balance which is maintained between the members of the various castes. Artificial removal of the royal couple results in the production of substitute sexual forms.

An interesting experiment has been performed by Castle and others with the soldiers. In the normal course, in a young colony only one larva will be transformed into a soldier. But if the single soldier is removed, another larva or nymph will quite soon be transformed. Moreover, the appearance of the first soldier can be suppressed by introducing one from another nest into the young colony. These results suggest that the presence of sexual forms or of soldiers *inhibits* the conversion of further nymphs into these castes, at least until the nymphs become very numerous, when the balance may be restored by the production of more individuals of the required caste.

In another curious experiment Grassé and Noirot made up isolated pairs of fourth stage nymphs of Calotermes. They found that if the pair was one of each sex both became sexually mature after the next moult. But if the two were of the same sex, only one became sexually mature—another example of balanced production of the required castes. In normal colonies of Calotermes, if the royal couple is removed, substitute sexual forms are produced, but only one pair becomes the royal couple and the others are destroyed by the workers. In some other termites, however, a number of substitute sexual forms may reproduce simultaneously, perhaps because their rate of egg-laying is lower than that of the original queen.

A final example of social regulation is provided by an experiment by Miller on the American Prorhinotermes. He put a group of 545 similar nymphs into an artificial nest and after several moults the following individuals were produced :

Large workers	Soldiers	Substitute sexual forms	Stayed as nymphs or died
103	37	82	323

This result illustrates another characteristic of termite development : that moulting can lead either to progressive or to retrogressive development. Thus the large workers and soldiers have actually lost the wing-pads which were present in the nymphs, whereas the heads and mandibles of the soldiers and the reproductive organs of the sexual forms show a positive change. The possibility, within limits, of developing either forwards or backwards gives the caste system great flexibility.

THEORIES OF CASTE FORMATION

It has been mentioned that both the original royal couple and the soldier seem to inhibit the production of forms like themselves, either temporarily or permanently. This observation led certain American students to propose the theory of caste control by " social hormones ". It was suggested, for instance, that the secretions of the royal couple, which are so eagerly licked up by the workers attending them, would be shared all through the colony by the mutual interchange of food, and might, like the effect of some drug, inhibit the development of sexual maturity in other members of the group. A similar theory might explain the inhibitory effect of the soldier, though in that caste the secretions to be handed round are much less copious.

Such theories meet two initial difficulties. First, in many termites the variety of castes would require a surprising number of special substances ; secondly, winged sexual forms are produced in colonies containing a royal couple. It is only the substitute sexual forms which are inhibited. The difference is that the substitutes would reproduce in the colony already containing the royal couple, whereas the winged forms would leave it to found new colonies elsewhere. The inhibition, therefore, is of a peculiar type.

Lüscher has given a good demonstration of social hormones on Calotermes flavicollis. The queen can be fixed so that her head is on one side of a screen and her abdomen and the king on the other. Sexual forms are produced in the part of the colony which has access to the queen's head, not in the other. Since a queen alone is only partly effective, both male and female hormones are probably required. These are swallowed by the nymphs and pass from one to another via the faeces. A nymph fixed in a screen, with its head on one side and its abdomen on the other, will serve as the channel through which the hormones pass from one half of a divided nest to the other. Castle and Light also obtained similar effects by adding extracts of royal couples to the food of isolated Zootermopsis nymphs.

The social hormone theory is but one of several " trophic " explanations of caste production in termites. Some authors have supposed that foods of the ordinary sort, together with varying amounts of salivary secretions, might control the process. At present it is very difficult to accept any such theory. The food varies considerably with the age and the condition of the colony. In young colonies, the sexual forms are largely reared on vegetable matter, and the same is true of substitute sexual forms at any time. There is no evidence for a copious supply of glandular food, such as is given to the queen-producing brood of the honey bee. Moreover, the self-differentiating group of nymphs scarcely seems to have the possibility of providing itself with the necessary variety of food.

The theory put forward by Grassé is that caste-determination is essentially a " group effect ", acting through the nervous system. This implies that the physical development and behaviour of a termite is determined in part by the sensory impressions which it receives from the other members of the colony. This might be automatically reinforced by differences in feeding which become established as the behaviour of different individuals diverged. Group effects of a kind are, of course, common in human beings. We learn the language and often the vocabulary and trade of our immediate neighbours. But this is the result of imitation, of learning and of the legacy of traditions, none of which is operative amongst termites. A closer but still false analogy is with the revolt, often subconscious, which children sometimes make against the ideas or way of life of their parents.

Group effects are, however, well known in many insects, though none is quite comparable with what is assumed to happen in termites. One example has already been described, namely the inhibition of oviposition in humble-bee and Polistes workers as long as the queen is present. In Polistes, the removal of the queen releases egg-laying within a day or two. Group effects have also been noted in various beetles and flies when bred in laboratory cultures. Grain weevils, for instance, do not lay at the highest rate of which they are capable unless each female has at least fifty grains of wheat, so that twenty females on six hundred grains lay less well than the same number on one thousand grains. Moreover, the weevils bred from a moderately crowded culture are larger and are capable of laying more eggs than are those reared from cultures which were either less or more crowded. This last effect seems to be due to physical changes in the culture such as the accumulation of the water given out during respiration. The effect of crowding on the rate of egg-laying is, however, likely to be due to nervous stimulation.

A closer analogy to what is presumed to happen in termites is found in the migratory locust. In this insect, both in nature and in laboratory cultures, the individual reared in a crowd is profoundly different from one reared in isolation. It takes on a different colour, the shape of the thorax becomes perceptibly different, the wings become longer and the behaviour is entirely altered. The nymphs tend to march about together in bands instead of hopping about as individuals, and the adults tend to undertake migrations in large swarms. It is certain that these differences are mainly due to crowding, and probable that they arise mainly from the sensory impressions received through the eyes. Living together in a crowd and seeing their fellows moving, makes each individual more active, and this in turn in some way alters the colour and shape during development. No doubt this is a much simpler process than those which would have to be invoked in the case of termites, since in the latter several different developmental paths are open and each path is chosen in approximately the right proportions. The theory of the group effect in termites cannot at the moment be regarded as more than an interesting hypothesis requiring further work.

GROWTH AND DIVISION OF COLONIES

The multiplication of termite colonies by the marriage flight of the winged adults, though the normal method, is not the only one. In some species, e.g. Calotermes, if the nest becomes very diffuse and spreads a long distance underground, the inhibition exercised by the royal couple may be no longer effective throughout, and substitute sexual forms may develop. Such outlying sections of the colony may eventually bud off, like the small bulbs found round a larger one. Much less commonly, in a few kinds of termites, more active splitting up of the colony has been observed, with all its members walking out of the nest, carrying the eggs and young larvae, and breaking up into groups, one with the original royal couple, the others with substitute sexual forms.

The ability to produce these substitutes means that a termite colony is potentially immortal, since its reproductive functions can always be renewed. It seems, however, that in many species substitute sexual forms do not play the important rôle which they do in laboratory experiments. Except in certain species, they seem to be relatively uncommon, probably coming into play when one of the royal couple is lost in some rare accident. There is some evidence that as the colony gets older, substitute forms are even less effective, so that in reality most termite nests probably last only a moderate number of years.

Calotermes colonies usually appear to survive for twelve to fifteen years. Other species in which substitute sexual forms are produced more regularly and in larger numbers may well last longer. The very large colonies of the higher termites must certainly be long-lived, if only because the huge nests must be constructed over a long period. Hill has recorded that the top of a large Australian steeple-like nest was knocked off in 1872 to allow a telegraph wire to pass over it and that the colony was still flourishing in 1935, when it must have been much more than sixty-three years old, perhaps nearly one hundred years. Apparently, the original royal couple are indeed able to live for a period as long as this, a much longer period than has been recorded in any other insect.

These old colonies may contain an enormous number of individuals. Thus in a South American species, with a nest measuring about 2 by 1 yards, Emerson found three million individuals. These large numbers can be arrived at by counting samples. Many of the large African nests certainly contain many millions of inhabitants. On the whole the nests of the less specialised termites are much smaller, having only a few thousand termites and flourishing for a much shorter period.

One interesting exception is the Australian Mastotermes, which as far as structure goes is the most primitive of all termites, but whose colonies may nevertheless contain several million individuals. It was mentioned earlier that the eggs of this termite are glued together like those of a cockroach. The insect too is cockroach-like in a number of details of bodily structure, and this strongly supports the view that it was from an insect of the cockroach-type that the termites were evolved in the remote past. Some of the species of cockroach which live out of doors show a certain resemblance to termites in the care which they devote to their young and in their wood-feeding habits.

Mastotermes itself seems to be the one surviving relic of a type of termite which was once found all over the world. Fossils closely resembling it have been found in the rocks in a number of places, including the Isle of Wight, which at the time the insects were alive must have had a much warmer climate than it has to-day.

This account of termite life has been built up from the observations and experiments of a large number of workers in all the continents, mostly during the last sixty years. The whole body of information makes an impressive picture of their strange social organisation. Yet one may well feel that our knowledge of them is still quite rudimentary, that there is far more to discover than has so far been revealed. We know least about the termites which make the largest and most perfectly organised nests. We know nothing about their powers of communicating with one another, which may well be as effective as in the honey bee. The great attention which has recently and is now being paid to them may lead to startling discoveries.

9

INSECT SOCIETIES

Insects, including even the most successful social species, are so unlike man that their resemblances to one another seem much more important than the differences. We feel, I think rightly, that even the solitary cat is more like us than the most socially-minded ant or the most docile termite.

Although the problems which have to be solved by any animal which becomes social have always something in common, the possible solutions depend mainly on the structure and behaviour which the species has previously acquired in the course of its evolution. The principal difficulties which have to be mastered are, first, how to control reproduction; second, how to obtain enough food from a limited territory; third, how to substitute co-operative and docile for solitary, aggressive behaviour; and fourth, how to adjust behaviour to the varying current needs of the community. The solutions of these four problems by the various types of social insect may be compared with what we find in man and to a less extent in other vertebrates. The problems are not, of course, independent of one another, though it is convenient to consider them separately. For instance, if the food supply were unlimited, control of reproduction would be less important.

Explanation of biological observations can often be made at several levels. Consider the fact that ants often eat a proportion of the eggs which they lay. The immediate cause of this behaviour in the individual ant may be merely hunger or more specifically, need for extra proteins. But the circumstances of the whole colony for several preceding weeks may have subjected that ant to protein starvation. Over a much longer period it may have been advantageous for the species to develop a habit of egg-eating when the rate of multiplication was out of balance with the food supply. In the discussion which follows, many of the arguments are chiefly concerned with the more remote levels. An individual insect may behave in a way which is advantageous to it or to its colony in the long run, without any suggestion that it knows what it is doing. Some behaviour, indeed, such as egg-eating, may seem harmful in the short run.

REPRODUCTION

While all species, including man, are capable of rapid multiplication if enough food is available, there are good grounds for thinking that the rate characteristic of each species has been subject to evolutionary change. A species which produces more young will not benefit unless more of them are also brought to maturity. Many marine animals, such as fish or molluscs, produce immense numbers of eggs of which very few survive. In many of these animals the eggs and sperm are merely liberated into the sea, and fertilisation is a haphazard process. A less wasteful method of pairing is a prerequisite of a smaller egg-production. Insects commonly produce a few hundred eggs each, and of these a higher but still small proportion reach maturity. It is advantageous to reduce the number of eggs only if at the same time they have more chance of survival, probably through the development of maternal care.

Most insects do, in fact, lay their eggs in some more or less protected situation on or near the food which their young will eat. The solitary bees and wasps, which lay still fewer eggs than most, look after them even more. Although no solitary ants or termites are known, we can be certain that the ants had solitary, wasp-like ancestors, to whose habits the colony-founding queen of most species reverts. The past history of the termites is less well understood, and it is possible that co-operation between males and females, rather as in the bark beetles, may have existed before they were really social.

The reduction in the egg-number of nest-making insects is a redistribution of physiological energy. If the female is to live longer and to work hard at making a nest, it is impossible for her to produce large numbers of eggs at the same time. It may be in a sense an accident of the past history of a species which determines whether it is capable of balancing a reduction in eggs by taking better care of them.

Nevertheless, some maternal care has appeared again and again in the course of evolution. Nest building occurs in various fish, such as the stickleback. Here the eggs are guarded by the male and the number laid by the female is very small. The land vertebrates, especially birds and mammals, nearly always produce fewer young than any solitary bee or wasp, but show a similar or higher degree of care for them. Most species restrict their breeding to a definite season of the year. Even in tropical countries there are usually seasonal differences in rainfall or in the quality of the light, and the breeding of most animals is adjusted to the seasons. Thus one very common restriction on reproduction is to confine it to a relatively short breeding season. This not only slows down multiplication but adjusts it to the most favourable time of the year.

Many birds, such as our common finches, aggregate into flocks in the autumn and winter, and separate into pairs only when they breed in the spring. This illustrates a difficulty in adjusting reproduction to social life more fundamental than the mere avoidance of over-eating the food supply. Sexual reproduction tends to have a disruptive effect on social organisation. In solitary bees and wasps one can often see the nest-building females being pestered by ardent males which still attempt to pair though the busy females no longer need their attentions. Man is almost the only social animal in which nearly all the adults of the right age are capable of breeding at any season of the year. Probably all the other social vertebrates, such as deer or beavers, have a relatively short breeding season, outside which the relations of the sexes are platonic. The majority of birds and mammals also have some sort of territorial system. The male robin, for example, at the beginning of the nesting season, occupies a territory out of which other males are driven. His song advertises his presence to unpaired females and also warns off other members of the species. In this way he claims an area of about the right size to supply one pair and their young with food. In most seasons, some of the males fail to obtain a territory and do not breed. Social insects, though they may, as in ants, only forage in a limited area, defend the territory only in the immediate neighbourhood of the nest.

The problem of too rapid multiplication in the social insects has been met in a different way. One of the first steps in the evolution of their social life was to establish a sterile caste which either did not breed at all or did so only to a limited extent and under certain conditions. They established a distinction between egg-laying to produce workers and egg-laying to produce sexual forms ; only the second process needs severe regulation. In the less advanced societies, such as those of the common wasps, the production of sexual forms leads to the break up of the colony. This does not seem to be merely a modification of behaviour to meet the difficulties of our winter climate, since colonies of some tropical species disperse after a comparable developmental cycle.

The termites, owing to the nature of their sex-determining mechanism, at all times produce males and females in equal numbers. It is probable that in them the sterilisation of most individuals of each sex must have developed very early in their history.

The Hymenoptera, the group of which ants, bees and wasps form a part, normally have a type of sex-determination which makes it possible for purely female broods to be produced for long periods. Once a female has mated the actual fertilisation of each egg is under nervous control by the female, and normally only unfertilised eggs will produce males. While this might be expected to lead to a colony rather different from that of the termites, the problems to be solved are much the same.

In solitary bees and wasps, unfertilised females do not seem to make nests, though there is still a large field here for experimental study. On the other hand, fertilised egg-laying females are commonly quarrelsome and tend to interfere with each other's egg-laying. Thus the sterilisation of most of the females without depriving them of the urge to work must in most species have occurred quite early. Moreover, since unfertilised eggs will develop into males, merely to deprive the female offspring of the queen mother of their opportunity to mate does not prevent reproduction.

In the colonies of Polistes wasps it seems that the prevention of egg-laying is still a mainly psychological inhibition. In the more advanced social species, the sterilisation of the workers has gone much further, though egg-laying can still occur, at least occasionally or in some individuals.

Sterilisation might be described as the least wasteful way

of preventing reproduction. Social insects do, however, also practise infanticide, the eating of eggs and larvae. The honey bee is peculiar in also destroying adult males in the prime of life as soon as they become superfluous.

While the less advanced insect societies break up as soon as the next generation of sexual forms has been produced, many species are able to emit huge swarms of males and females without disrupting the colony. Ultimately, this seems to depend on evolving a sufficiently long-lived queen or royal couple around whom the workers are organised. The break-up of the colony is probably more associated with a decline in the oviposition rate than with the actual appearance of the reproductives. In the termites and the honey bee the production of substitute sexual forms gives the colony further stability. Some ants may achieve the same end by receiving back some of the young fertilised queens.

It is the peculiar triumph of man to have established social communities more elaborate than those of ants and termites without sterilising the majority of the adults. Other social mammals have a relatively simple community and only a short breeding season. In some primitive human societies infanticide, especially of females, has been practised. In all our societies a great variety of taboos and customs tends to discourage promiscuity and partially to reduce the birth-rate. Until very recently a high infant mortality and periodic famine were the final but wasteful agents by which the population was kept in equilibrium with the food supply. If the progress of medicine and agriculture save us from these, we shall presumably have to rely on family limitation.

It is no accident that the queens (or in termites, the royal couple) of social insects live much longer than the females of most solitary species. The full development of maternal care demands a wide overlap between successive generations. For species which are not social, this overlap need only last until the next generation is self-supporting. In man, social life has developed through the learning of crafts and the handing on of traditions. Thus the great length of childhood in man compared with any other animal has both made a long period available for education and has required, in compensation, a relatively long life for the adult.

Insects, in contrast, seem to depend very little on learning. Their elaborate nests, the hunting territories, and some of their food supplies may, like human traditions, be handed on to the next generation, but would be quite insufficient in themselves to preserve a stable society. The fundamental reason for the greatly extended life of the queen is that she stabilises the whole system by means of her oviposition. Apart from periods of cold weather, no insect society is in a stable condition unless the queen is laying regularly. The whole system of inhibiting unwanted reproduction depends on her, so that the colony usually survives very little longer than the queen. The colonies of termites and of the honey bee which no longer depend on the survival of the original queen are potentially immortal.

FOOD AND SOCIAL LIFE

The importance of food in the early stages of social evolution may have been much less in insects than it was in man and some of the other social mammals. In early man and in such animals as wolves, co-operation in hunting must have been one of the original social motives.

In the social insects co-operation for this purpose is only found in the ants, and in them chiefly in the more specialised types. Wasps never hunt in packs, and bees forage alone, though in the honey bee workers may tell one another where to go. Termites are only rarely hunters and most of them eat the same sort of vegetable food as probably satisfied their solitary ancestors. This was at first made possible by their special relations with the protozoa in their intestines. Similar intestinal inhabitants are known in some solitary, wood-feeding cockroaches. If this specialisation is lost, a more varied diet at once becomes necessary.

The disadvantages of living by hunting other animals are that the supply is too variable and not really large enough to allow many hunters to live in one place. The driver and legionary ants have carried this type of existence to its extreme development, at the expense of being permanently nomadic. They appear to be a successful sideline from the main stream of ant evolution. Bees and wasps have attained to a different solution. They have never developed any private source of food, but compete with a large number of other animals. It is true that bees and flowering plants have had a profound influence on one another's evolution, but nectaries which are open to bees but closed to other insects are possible only to a limited extent.

The wasps are still mainly hunters, and only a very few of them, such as a few of the South American Polybiines, compete seriously for nectar. Bees and wasps have, however, developed great powers of flight and an ability to work hard. While honey-bee workers will get their nectar and pollen from near at hand if they can, it is not at all unusual for them to fly 2 miles or even considerably farther.

It is well known that in man large communities were possible only after agriculture had been developed. This not only provides much more food per acre than hunting does, but also supplies it in a form in which it can be stored for use in winter. The honey bee and the honey-gathering wasps, although not agriculturists, both lay up similar reserves. The termites seem on the whole to manage by feeding on abundant widely available substances for which competition is not very keen, since special modifications are needed if they are to be utilised economically. Only the most advanced families cultivate fungi, and the value of this crop as food is uncertain. Most termites, however, can and do lay up food-reserves.

Apart from man, it is the ants which have developed the most varied and also probably the best solutions of this problem. While the more primitive types still live almost entirely by hunting other insects, the others have exploited a gradually widening range of foods. This has been associated with a great increase in the size of the colony. The greater availability of food may, however, be only a part of the reason for this.

The exact relations between ants and their green-fly cattle are in doubt, but it is certain that they have much less control of their beasts than we have of ours. Cattle by themselves, without an agricultural industry which provides feeding-stuffs, do not much increase the yield per acre.

This is a point which it would be interesting to study in such species as the British yellow mound-making ant, Lasius flavus, which is almost completely dependent on root-aphides. The cattle are suppliers of sugar and proteins, so that the ants have access to the riches of the plant kingdom without needing the special digestive arrangements of the termites.

Many ants also feed on plant seeds, and the harvesting ants do so extensively; but here again no great yield per acre is possible without a proper agricultural system. This has been developed only in the leaf-cutter ants with their fungus gardens, which they providently manure with their own excreta. It is probably not a coincidence that some of these ants make the largest nests with the most numerous inhabitants of any member of their family. It is not yet possible to say whether their solution of the problem of food-supply has enabled them to make other cultural advances. Our knowledge of their behaviour is still in a rudimentary state.

The general conclusion is that social insects have increased the variety of their diet, that is they use a greater proportion of what is available near at hand. Apparently very few have been able to increase the quantity of any one kind to be found within reach of the nest. It is possible that some of the ants which keep cattle have done so, and it is almost certain that the leaf-cutter ants have solved this problem.

THE ORIGIN OF SOCIAL BEHAVIOUR

The third point to be considered is the acquirement of social behaviour by species which were originally solitary. This also involves the idea of division of labour, and ulti-mately the question of communication.

The simplest example of the division of labour is seen where two sexes share the duties of reproduction. There is a number of animals which show us that such a division need not be made. In many molluscs, such as our common land-snails, every individual performs the function of both sexes. Nevertheless, in the great majority of species the sexes are separate. This applies to almost all insects, and to all verte-brates.

It is characteristic of insect societies that the division of labour amongst their members is founded chiefly on physical differences which are often well marked. Apart from differences in behaviour which depend upon age, all the individuals which have distinct rôles in the society differ from one another in structure. The differences between the queen and the worker, between the soldier, the worker major and the worker minor, are often very marked and quite as big as the differences between species of solitary insects.

If workers have a uniform structure, then at different times they contribute to all the activities of the colony. The castes of social wasps are feebly differentiated, and in them permanent division of labour, except between the queen and her workers in the mature colony, scarcely exists. In the termites, by way of contrast, we find the worker and the soldier often with a strikingly different appearance and with quite distinct functions. Even more curious is the example of the amazon ants, in which the amazons are in effect the soldiers and the worker caste is represented by slaves of another species. It seems that the essential nature of an insect society must be looked for in the physiology of the individuals.

Before pursuing this point further, we may notice how different has been the evolution of social vertebrates. Structural and physiological differences within the species are relatively trivial apart from the differences between the sexes. Perhaps because of this, only man has been able to build societies of a complexity even approaching that of the ant or termite colony. Without physiological differentiation, it is very difficult to organise a society until there are elaborate means of communication and a brain capable of understanding what is conveyed.

There are a few exceptional animals like the beaver about which one would like to know much more, but in general the non-human social mammals combine only in a rather rudimentary way, either for hunting or for defence. Any division of labour depends either on age or on sex; for example, bulls defending the cows and young amongst wild oxen. It might be argued that in man there is a division of labour based on innate differences in abilities, but there is very little physiological differentiation. The superficial human variation in stature and in colour of hair and skin are not related to social duties. With the requisite training, most men can do most of the essential work of the community.

To return to insect societies, clearly the fundamental problem of their origin is, "How did physiological differences arise within the species?" Solitary bees and wasps construct nests, and only a slight lengthening of the life of the mother would be enough to make it overlap those of her offspring. They are also capable of producing exclusively female broods, though none of the solitary species is known to do this except occasionally by chance. Such Hymenoptera, therefore, could at once set up a primitive society of the Halictus or Polistes type if they could produce a brood of young females who would work without laying eggs.

There is little doubt that an important factor has been the modifying influence which social life has on the process of natural selection. In a solitary species, individuals with reduced fertility will not often survive in competition with others which are more fertile. But in social species any change which benefits the group as a whole is likely to be preserved.

In insects generally a reduction in size of the female is nearly always associated with the laying of a smaller number of eggs. It is significant that in most species the worker is smaller than the queen. It is possible that originally, as is still true to some extent, queens and workers differed chiefly because of chance variations in the amount of food which they received as larvae. Later, a system could have been evolved in which differences in the food received had a greater effect on the adult, whereas in solitary species the tendency would be to keep the species as insensitive as possible to variations in food supply. When physiological studies have been made of the effects of nutrition on the size, structure and fertility of Hymenoptera it may be possible to argue the case in greater detail.

The beginning of termite societies was probably quite different. There are several kinds of insects, such as the Embiids, in which all the developmental stages are known to live together, sharing a common silken shelter but not helping one another in any other way. What seems to have happened in the termites was that the developmental stages began to play a part in the life of the community as important as that of the adults. This was possible in them because of their gradual metamorphosis and the general similarity of the early stages to the adult. The caste system of the termites would originally have been founded exclusively on age, the half-grown larvae or nymphs doing most of the work while the queen laid the eggs. This system would increase the chances of variations in nutrition during the developmental period. As in the Hymenoptera, an increased response to the effects of nutrition might have been evolved later. Such a scheme would give at least a hypothetical explanation of the origin of the more primitive termite societies. We still know so little of the complex caste system of the higher termites that any discussion of them is premature.

INSECT SOCIETIES AS FAMILIES

An insect society is a family, a group of offspring surrounding the mother or royal couple. Their mutual tolerance is founded on growing up in the same nest. If ant colonies are split into two in artificial nests, the two halves may fight if reunited after an interval of some months. Judging by ants and honey bees, mutual tolerance depends on sharing a common nest-smell, in natural conditions acquired during development. Ants of the same species but from different colonies fight bitterly, and colonies of the honey bee can be combined in one hive only by devices which enable them to acquire a common smell. It is very unusual, perhaps even unknown, for separate colonies to combine in natural conditions.

Thus while the insects have species and families or colonies, they do not have tribes or nations. The failure to develop co-operation on the larger scale which occurs in man may be partly due to their much higher fertility. In man, several families had to combine to form a group big enough to master a mammoth, let alone build a cathedral. But insect families are so much larger that there would never have been the same advantage in combining with strangers.

When two insects meet they do not normally fight unless one forms the natural food of the other. What might be called nasty aggressive behaviour, worthy to be compared with our own, is always in essence the fight for an oviposition-site. Some flies which lay their eggs in fruit will knock other females off in order to get their place on the fruit, but aggression is usually found only in females with a nest to defend, and they will drive off any other insect which happens to come too close.

Members of one colony would, therefore, live together amicably if they all smelt the same and if there were only one ovipositing female. In itself, this would not lead to much in the way of co-operation, and probably at first the nest would be an aggregation of the single nests of each individual. Something like this seems to be still true of Halictus. But if one started with an insect which already had behaviour as elaborate as that of the solitary Ammophila, it is reasonable to think that the great advantages of co-operation within the family group would provide plenty of opportunities for selection. A great deal of the social co-operation of insects is no more than the accomplishment of the same task by individuals which are all in the same physiological state. It is only in the most advanced species, in ants like Oecophylla, in the honey bee, and in the specialised termites that cooperative behaviour can clearly be seen.

It may seem far-fetched to suggest that all ethical systems are ultimately founded on the necessities of social life. Something of course depends on the definition of the word ethical ; if it is restricted to behaviour determined by a conscious plan of action, all animals other than man would seem to be excluded. It is perhaps more reasonable to regard the refined forms of altruism which we consider the highest type of ethical behaviour as developments from the maternal care of many solitary animals, and especially from the family loyalty of social species. If a bird defends her nest against an enemy, her behaviour is often said to be instinctive or innate and therefore no more creditable than eating a worm. In the social insects, workers will defend their nest to the death, and will care for young which may be only remotely related to them. In human beings, on some occasions and by some individuals, the whole species is treated as a family to be defended or cherished. Probably very little of our altruism is instinctive or even unconscious. Nevertheless, the instinctive altruism of the primitive family may have made the pattern from which the traditions of conscious altruism were later evolved.

Our assessment of ethical values is associated with a conscience, a sense of right and wrong. In its lowest form this is a desire to win the approval or to avoid the disapproval of the rest of the community. Domesticated and therefore semi-social animals like the dog can be trained until they have a rudimentary conscience, founded on an experience of rewards and punishments.

The process of natural evolution can be looked on, by analogy, as a slow but very prolonged process of training in which the rewards are food and successful reproduction and the punishments hunger and extinction. In a social mammal or insect, other members of the community are an essential part of the environment without which the individual is almost as helpless as a fish out of water. This reaction to other individuals is innate and is the only element of social behaviour which is common to all such species.

It cannot be supposed that social insects have any sense of right or wrong. So far as they are obeying the laws of their own physiology, they are very unlikely to have moral scruples. But from the evolutionary point of view motives are unimportant; ethical conduct is what benefits the community, even at the expense of the individual life. The code developed during evolution, because it is such a slow growth, is apt to be too rigid to succeed on all occasions. It leads to those "errors" of instinctive behaviour which Fabre and some other naturalists have treated as contradictions of the Darwinian theory. This, however, is a misapprehension; natural selection is seen just as clearly in the failure as in the success of a behaviour pattern, provided that in the long run the results determine survival.

The adjustment of individual behaviour to social needs is closely related to the topics already discussed. An insect living in a society has to do a greater variety of tasks, at least until an advanced division of labour has been evolved, than does a solitary species. Its action cannot follow any very rigid time-table. Perhaps the various methods used in cooling the nest when it gets too hot are the most striking example of sudden adjustment of behaviour to the needs of the colony. If a solitary species gets too hot, it may try to escape, but it has never been known to do anything to make its nest cooler.

Although reactions to novel situations have not been much studied, they are clearly better developed in the more specialised social species.

Most wasps make no use of old nests, even in the tropics where one can often see deserted ones in good condition. It is a curious fact that whereas various solitary wasps will build mud cells in the old combs of Polistes nests, the Polistes themselves never seem to use an old nest, though they may occasionally suspend a new one from it. The honey bee, which will use a machine-made wax foundation as the basis for a new comb, shows a higher form of behaviour. Ants and termites, probably because they practice a less rigid type of architecture, seem to be much more adaptable. This is shown in the ease with which they can be domesticated in artificial nests. It is, however, in the rigidity of their behaviour that insect societies are most different and inferior to ours. The adjustments which they can make to an unfavourable or varying environment are relatively trivial. This rigidity is one of the basic differences between the insect and the mammal.

Even such mammals as rats and voles explore their habitat apart from an immediate hunt for food. One of the reasons why they get trapped is that they tend after a time to examine any new object placed in their territory. Exploratory behaviour is much more developed in monkeys, and is carried still further in man. This sort of activity has allowed man to develop complex societies without the physiological differentiation which was essential in the social insects. We have to pass education acts to prepare our citizens for social life, whereas insects are, to a very large extent, hatched already educated.

There is another point of some importance. Man has been social for, probably, less than a million years, whereas the higher social insects have lived in complex societies for at least thirty times as long. Few biologists would care to predict where we shall be when thirty million more years have passed. H. G. Wells, in his story *The Time Machine*, forecasts one possible result of such a period of evolution, assuming that we have a somewhat ant-like future. Such a course seems much less likely now than it may have done fifty years ago.

The efficiency of some of the different types of ant societies can be judged by their success when in competition with one another. The Argentine ant, a native of the southern part of South America which has been accidentally introduced into many parts of the world, is a good example, as Haskins has shown. It is a small, fragile species with a soft, easily damaged skeleton. The workers have no sting, but they can produce a sticky, scented, anal secretion which may be of some defensive value. It has now established itself in many of the warmer parts of the world and in some, such as the island of Madeira, it has completely eliminated the local ant fauna. Though it cannot tolerate cool climates like ours, even in the British Isles it has several times survived for some years in the shelter of buildings.

It seems to have two main advantages. First, the workers forage in large groups which are capable of an unusual degree of co-operation. Secondly, it has a specialised type of reproduction. The queens are small, hardly larger than the workers, but there are many of them in each colony : they have solved the difficulty of getting several egg-laying females to tolerate one another. This means that it is very difficult to destroy a colony, and that the queens are cheap to produce. This type of organisation is very successful, but not very mobile. The queens cannot found new colonies unaided and the colonies reproduce by " budding ", the splitting off of groups of queens and workers. This is the chief reason why the species did not spread until human transport was available to carry it about. But once this was provided its other superiorities took effect. When it is not given such transport, the other ants whose queens may fly long distances to found new colonies have a compensating advantage, at any rate outside small islands.

THE STUDY OF SOCIAL INSECTS IN THE FUTURE

The progress of biology in the last hundred years has been associated with the gradual application of the experimental method to more and more varied fields. The experimental study of insect societies is quite recent. A beginning was made in the nineteenth century, notably in the study of Italian termites in the 1890s, by Grassi and Sandias. But most of the tools necessary for such a study have become available only recently. This is especially true of the basic facts of insect physiology and of the methods and ideas used in analysing the simpler behaviour of solitary species.

We can be sure that the experimental study of the organisation of social insects will lead to many more strange and even startling discoveries. For a long time to come, clear and imaginative thinking will be quite as useful as elaborate apparatus. A beehive or an ant-colony is easier to keep than a dog, and although it will hardly be so satisfying to the affections, it will provide a host of unsolved problems on which anyone may exercise his disinterested curiosity. It is the application of this distinctively human faculty, not only to insect societies but also to our own, which is most needed in the world to-day.

FURTHER READING

BUTLER, C. G. *The Honey Bee.* Oxford University Press, Oxford.

BUTLER, C. G., and FREE, J. B. 1959. *Bumblebees.* Collins, The New Naturalist, London.

DUNCAN, C. D. 1939. *A Contribution to the Biology of North American Vespine Wasps.* Stanford University Publications, Biological Sciences, 8: 272 pp.

EVANS, H. E. 1958. *The Evolution of Social Life in Wasps.* Proc. Tenth Internat. Congr. Ent. (1956, 2: 449-458).

FRISCH, K. VON. 1950. *Bees. Their Vision, Chemical Senses, and Language.* Cornell University Press, Ithaca, N.Y.

MICHENER, C. D. and M. H. 1951. *American Social Insects.* D. van Nostrand Co., Inc., New York.

SCHWARZ, H. F. 1948. *Stingless Bees (Meliponidae) of the Western Hemisphere.* Bull. American Mus. Nat. Hist., 90: xvii. + 546 pp.

SLADEN, F. W. L. 1912. *The Humble-bee, Its Life-history and How to Domesticate It.* Macmillan, London.

SNYDER, T. E. 1935. *Our Enemy the Termite.* Comstock Publishing Co., Ithaca, N.Y. (The most recent full account of termites is in French and by P. P. Grassé, in Traité de Zoologie, vol. 9. Masson et Cie., Paris, 1949).

WHEELER, W. M. 1913. *Ants.* Columbia University Press, N.Y.

WHEELER, W. M. 1923. *Social Life among the Insects.* Constable, London.

INDEX

INDEX

A

Acacia, ants' nests in, 136
Adlerz, 31, 41
Aerodromes, termite nests on, 178
Aggressive behaviour in insects, 204
Amazon ant: *see* Polyergus
Ambrosia beetles, 37
Ammophila pubescens: way-finding, 31; nesting, 40
Anergates, 165–6
Anoplius fuscus, 31
Ant thrushes, 125
Antennae, function of, 19
Anthophora, 76
Ants: and aphides, 121–2, 129; castes of, 119, 139–46; enemies of termites, 125; food of, 122; founding of colonies, 137; geological history, 120; guests, 151–2; nests, 134; sting, 125
Aphides, 121
Apis (honey bee), species of, 103
Argentine ant, 208
Atta, 132
Auxiliaries in Polistes, 62

B

Baerends, 40
Balm, 110
Bark beetles, 37
Beaumont, de, 153
Bees: and flowers, 77–8; development of social behaviour, 78–9; solitary, 76
Beirne, 56
Bischoff, 156
Black ant, common: *see* Lasius niger
Blewett, 21
Blood-red ant: *see* Formica sanguinea

B (continued)

Blue butterflies: caterpillars in ants' nests, 152; eaten by wasps, 50
Bols, 85
Bombus: *see* Humble bee
Bombus hortorum, nest census, 91
Bombus lucorum, 156–7
Bombus terrestris, 156–7
Bothriomyrmex decapitans, 158
Brachygastra, 129
Brain of insects, 20
Brauns, 82
Brian, 87, 92, 138, 142
Budding by ant colonies, 138
Bulldog ant: *see* Myrmecia
Burying beetle, 36

C

Cabbage white butterfly, 33
Calotermes, 183, 187, 188
Cannibalism in termites, 54–5, 175
Carpet beetle, 24
Carthy, 148–9
Carton made by ants, 135
Cassique, 69
Castes: bees (Halictus), 64–5; bees (honey), 105; bees (humble), 90, 92; bees (stingless), 99–102; termites, 179–84; wasps (Polistes), 64–5; wasps (Polybia), 71; wasps (Vespula), 45, 53
Castes, determination of: ants, 140–6; honey bee, 107; stingless bees, 100–1; termites, 184–7
Castle, 183, 185
Cellulose, digestion of, 176
Chambers, 77
Cinnabar moth, 33
Clausen, 121
Cleveland, 176
Clover: *see* Red clover

215

INDEX

INDEX

Honey bee—*continued*
determination of castes, 107; distance flown, 110; division of labour, 111; drones, 106–7; language and dances, 112–17; nuptial flight, 106; relation to flowers, 109–10; relatives of, 103; substances used by, 110; swarming and supersedure, 108
Honey-pot ants, 129
Hornet: in Cyprus, 56; effect of heating nest, 51; in England, 57
Humble bees, (Bombus), 83; cuckoo, *see* Psithyrus; and flowers, 93–5; hibernation, 84–5, 92; nest site, 83–4; nest size, 88; pairing, 87; sting 83; tropical, 95
Humboldt, 178
Hunting wasps, 40

I

Ichneumon wasps, 34, 42
Jhering, von, 66
Insight learning, 31
Instinctive behaviour, 25
Intersex, 144

J

Jet black ant: *see* Lasius fuliginosus

L

Labauchena, 158
Language: *see* Dances and Communication
Larva: definition, 24; in termites, 180
Lasius flavus (common yellow ant), 120, 199; variation in worker size, 139
Lasius fuliginosus (jet-black ant), 148
Lasius niger (common black ant), 148
Leaf-cutting ants, 132–44, 199
Learning, 27; unimportance of, in insects, 197
Legionary ants, 125; nature of queen, 139
Leptogenys, 122
Leptothorax, 142
Light, 185
Lobopelta, 125
Locust: *see* Migratory locusts

Longevity of queen: essential to social life, 196; in termites, 188
Lubbock, 122

M

Macgregor, 149
Mason bee, 76
Mass provisioning, 44
Mastotermes, 179, 189
Maternal care, 34, 193
McCook, 132
Mealybugs, 121
Melipona, 96
Membracids, 129
Metamorphosis, 24
Migratory locusts, effects of crowding, 187
Miller, 183
Monkshood, 93
Moulting: in insects, 22; in termites, 182
Myrmecia (bull-dog ant), 125
Myrmica (red ant), 128, 130, 137; scent trails, 149
Myrmica rubra, 138, 142

N

Nasute termite, 181
Nectar: collection by honey bee, 109; collection by wasps, 63, 70; exploitation of, 198
Newman, 84
Nixon, 56, 156
Noirot, 183
Noll, 81
Norwegian wasp, 56
Nuptial flight: of honey bee, 99; of humble bees, 89; of stingless bees, 99; of termites, 173
Nuptial promenade, 173
Nymph: definition, 24; of termites, 180

O

Oecophylla, 121, 124, 135–6
Okland, 124
Osmia: *see* Mason bee
Oviposition, provoking jealousy in queen, 38

217

INDEX

INDEX